ROBIN HOOD MATH

ROBIN HOOD MATH

Take Control of the Algorithms That Run Your Life

NOAH GIANSIRACUSA

RIVERHEAD BOOKS

NEW YORK

2025

RIVERHEAD BOOKS
An imprint of Penguin Random House LLC
1745 Broadway, New York, NY 10019
penguinrandomhouse.com

Book design by Daniel Lagin

LIBRARY OF CONGRESS CONTROL NUMBER: 2024051247

ISBN 9780593717875 (hardcover)
ISBN 9780593717899 (ebook)
ISBN 9798217176731 (international edition)

Printed in the United States of America
1st Printing

The authorized representative in the EU for product safety and compliance is
Penguin Random House Ireland, Morrison Chambers, 32 Nassau Street,
Dublin D02 YH68, Ireland, https://eu-contact.penguin.ie.

To my uncle Paul, who taught me about numbers and about life.

And to Jim Morrow, whose love for math was contagious.

Contents

ROBIN HOOD MATH

1

A Day in the Life
of a Number

Apiercing sound awakens you: the alarm on your phone. Through squinted eyes you peer in its direction and discern a blinking 7:00 a.m. This number denotes the beginning of your day—a day in which you will feel like the protagonist of Kafka's *Metamorphosis*, except rather than being transformed into a giant insect, you are reduced to an insignificant number. It's equally dehumanizing.

You reach to silence the alarm. Phone in hand but not ready to escape the ensnaring warmth of bed, you instinctively tap the TikTok icon to delay facing reality for a few minutes. Of the billions of videos you could be shown, TikTok's algorithm mysteriously chooses just a single video to show you first. Then another. Then one more after that. It uses a secret math formula to create this personalized playlist, but you have no idea how it works.

While you're mindlessly watching TikToks, a witticism strikes you, one you'd like to share with the world. You head over to Threads or Bluesky or X to do so but are confronted with a harsh realization: every remark on the platform, including the one you are about to post, is boiled down to a handful of numbers—views, Likes, reposts—that are on display for all to see. What if your witticism wallows in obscurity? Virality is a numbers game, but frustratingly you're not told the rules.

By now your mind has moved on to another thought: follower counts on social media compress your importance to a single number judged by humans and algorithms alike. Want to be a writer? Be prepared to tell publishers how many TikTok followers you have. Want to sell handcrafted soaps? The number of Instagram followers you have is likely more important than the number of weekends you spend at the local farmer's market. I've seen applicants for a teaching position beat out others for an interview by having a larger YouTube following. Even the importance of politicians is now sometimes measured by the number of social media followers they have. Real impact, meaningful personal connections, and deeper forms of friendship all become secondary to this shallow game of online quantification. Social media is pushing us all to act like social influencers—whether we want to or not. Time to put the phone down and start your day.

First up, a trip to the grocery store. You hop in your car, blissfully unaware that doing so results in numerification—the assignment of numbers to complex human acts. There's an entire industry, projected to be worth half a trillion dollars by 2030, dedicated to collecting and selling the trails of data your car emits when it hits the road. Everything is tracked. Your location and speed. Your acceleration and braking patterns. The music you listen to along the way. Who you call from your car and for how long you speak. Many cars today produce around twenty-five gigabytes of data per hour. That's roughly the size of all English-language Wikipedia articles—*every hour.*

While this data is used for safety features and navigation-assistance programs, it also fuels a range of businesses, from car manufacturers to insurance companies to tech firms, that want to know as much as they can about their customers so they can squeeze every ounce of profit out of them. And as cars are increasingly wired with internet apps—giving new meaning to the term "mobile device"—the auto industry is taking on the tactics of the tech industry. You know those annoying ads you see while riding in some taxis? It might not be long before they spread to private cars. And when they do, expect those ads to be microtargeted with your personal data just like all the online ads you see on your phone and computer. The old adage in marketing to know your customers has unsettlingly been taken to new heights in today's digital economy where everything you do is tracked, packaged as data, and sold to the highest bidder.

Fifteen minutes (and a quarter-Wikipedia worth of data) later, you arrive at the grocery store. Kroger, the largest grocery chain in the US, has over thirty-five petabytes of customer data. That's over 50 percent more data than the entire digital collection of the US Library of Congress. It uses this data to optimize its inventory and marketing campaigns—but it also sells some of this data to other companies in a lucrative data market. In the UK, Tesco and Sainsbury's sell customer data for an estimated £300 million a year. To put it bluntly, these stores don't just sell milk and eggs and other food products; they sell whatever private details they can collect from you.

You return home (after producing another quarter-Wikipedia of data on the drive), open your laptop, and decide it's time to face the painful task you've been putting off: applying for jobs. You search for openings in your area and get lots of hits. But something doesn't seem right. As you scroll down the page, you see link after link of crappy recruiter sites charging exorbitant fees, interspersed with job ads that aren't great matches for you. That's because the links you see don't represent the most relevant opportunities; they represent the companies that paid Google the most to put their ads at the top.

You might think of Google as a search company, but 80 percent of its quarter-trillion-dollar annual revenue comes from ads—both hosting them and placing them throughout the internet. And a big part of what makes Google the most profitable advertising company in the world is that it knows a *lot* about you. This includes what you search for, when you search for it, how often you click different types of links, what videos you watch on YouTube (which is owned by Google) and how long you watch them, and what you type in your Google Docs and Gmail if you use those apps. All your actions online are fuel for the ad-targeting industry. Google and other tech giants now have another fancy tool for turning you into a pile of numbers and personal data: chatbots. These aren't just friendly personal assistants; they're data harvesting machines that give their developers an even more intimate algorithmic window into your thoughts and desires.

Back to your job search. You swim through the sea of sponsored content and find a few jobs worth applying to. Large companies today typically use computer programs to handle the mass of applications they receive. Some rate how well applicants match the positions being applied to, and some rank applicants to save recruiters the time and trouble of doing so manually. You are once again reduced to a number. It would be nice to know what keywords you should include, and what experiences to highlight, to increase your odds of receiving a favorable number. But don't expect the cold, calculating algorithms to provide you with any helpful feedback. They are programmed to maximize efficiency, not humanity.

If you're lucky enough to land a job, prepare to be treated as a number there, too. From truck drivers and factory workers to white-collar office workers, employees of all types are increasingly monitored by productivity tracking apps—tying hourly pay, bonuses, and continued employment to a laundry list of surreptitiously recorded statistics. Of the ten largest private employers in the US, eight now use this technology in some capacity. Some office workers recommend purchasing a mouse-jiggling device that keeps the cursor moving so you're not dinged for idle time.

At an investment management firm in Connecticut, each employee is scored on seventy-seven different measures. These performance numbers are displayed like batting averages and slugging percentages on a baseball card. Perhaps this focus on metrics works: this firm, responsible for over $100 billion, is the second-most-successful hedge fund on the planet. In Minneapolis, a hospice chaplain's employer tallies "productivity points" of a different kind. Visiting a dying patient is worth one point; participating in a funeral is worth one and three-quarters. From 2019 to 2022, as the pandemic pushed many into remote work, global demand for employee monitoring software rose by 65 percent. A legal advice firm in the UK now coaches businesses on the pros and cons of using AI to "track employees' performance, monitor their actions, and predict their future behaviour"— something we'll surely see a lot more of in the coming years as AI continues to proliferate.

Uncomfortable with this relentless drive for efficiency, you decide to take matters into your own hands. If you start your own business, you'll be your own boss and escape the quantitative confines of productivity trackers. First stop: your local bank, to apply for a business loan. Alas, you are once again a number as your credit score dictates your ability to secure funding.

Perhaps it is instead time to go back to college. Standardized tests ensure you're a number before you apply, while college rankings affix a number to your degree after you graduate. Think your time in between will be less of a numerical nightmare? Think again. Many schools issue students official laptops that automatically report to professors how much time each student spent on the required readings and how long they looked away from the screen during an exam.

The tracking software installed on these laptops also reports data to campus police and college administrators—and to private companies that leverage students' intimate information for profit. The parent company of the learning management system Canvas, used by over a third of higher-ed institutions in North America, once said that its troves of education-related data are key to the company's multibillion-dollar valuation. Cengage, a massive online textbook company, doesn't just track how many pages each student reads. It also collects information on webpages visited, links clicked, keys typed, and more. And it sells portions of this data to companies that use this information to target students with invasive ads.

Seeing all the ways in which you are treated as a number is enough to make a person sick. Speaking of being sick, don't expect a break from this incessant numerification when you go to the doctor. And I'm not talking about numbers like heart rate and blood pressure. Health insurance companies plug you into their formulas to compute your premiums and payouts, hospital managers use formulas to shape staff sizes and hours, and triage nurses use a point system to determine your spot in line at the emergency room. Need a kidney or liver? There's a number for that too: algorithms rank potential recipients to determine who gets the next available donated organ.

At this point, it might feel like your life itself is just a number. To some government agencies, it literally is. FEMA put the "value of a statistical life" at $7.5 million. The EPA is more generous, using a $10 million price tag. Well, that's for lives in the US. In at least one document trying to assess the cost of climate change in terms of lives lost, the EPA controversially decided to adjust the value of a statistical life based on per capita income. This means the number assigned to someone in Germany is ten times the number assigned to someone in Ghana.

Those in power have long been inclined to see people as numbers. This partly reflects a practical constraint. You can treat the members of your family and your local community with the individuality and respect they deserve. But a government is responsible for so many citizens, and a large company for so many customers, that individual faces and identities inevitably blur into a fog of impersonal statistics. However, it goes far beyond this. Treating people as numbers is also a way to uncover patterns in behavior and optimize operations for efficiency. That's because numbers are the language of mathematics, and mathematics provides an array of potent tools for analyzing trends and making predictions.

While numerification is not new, its extent today is mind-boggling. The apps on our phones and computers and in our cars have made our lives more comfortable and our activities more convenient, but they have also enabled the collection of a staggering amount of personal data. This data turns us all into numerical grist for the mathematical mill, allowing the powers that be to predict, optimize, and profit with unprecedented scale and success.

What can we do to regain our humanity and autonomy? I believe we must reclaim the mathematical tools currently being used to manipulate and exploit us by the public and private sectors alike. Math is commonly portrayed as something only prodigies and savants can do. But this is a myth. As I hope to show in this book, you don't need to be an algebra all-star or geometry guru to avail yourself of the wonderous benefits of math. You just need math to be presented in a new and more inclusive light, focusing not only on the formulas that techies and quants use to get ahead—but also on how ordinary people can use these formulas to confront the challenges faced in everyday life. I call this Robin Hood math because it takes the power of math from the rich and gives it to the poor. This won't stop you from being seen as a number, but it will help bridge the chasm that has grown between those with quantitative backgrounds and those without. To preview how this works, let's revisit the Kafkaesque day you just experienced.

Rewind

You start your day with a few TikToks in bed, but this time armed with knowledge of the app's secret formula. In chapter 7 you'll learn what the formula is, how it escaped the clutches of the tech giants, and how you can use it to improve your social media experience. Suddenly all the tricks content creators use to nudge their videos into virality leap out at you. And rather than taking a passive role on the platform, you choose your actions more carefully to cultivate a feed you feel good about. The entertaining spontaneity of an endless stream of videos is still there, but the confusion and frustration you used to experience have largely abated. Feeling inspired, you create and post your own video—and you do so strategically, with a mind toward the numbers the secret formula looks for. You are rewarded with the views that have long eluded you. Time to put the phone down and start your day.

First up, some groceries. At the store, you recognize the tell-tale signs of data-driven management. Grocery stores in wealthier neighborhoods put the produce section near the entrance while in less affluent neighborhoods it's relegated to the back and snacks and baked goods are at the front—because the data shows more money is made this way. And the data helps store managers determine which items are trending so they can be placed at the checkout as impulse purchases. As with clickbait on social media, knowing the math behind these manipulations helps you resist them. To an extent, at least: those dark chocolate–covered almonds still found their way into your cart at the last moment.

Once home, you brace yourself for the daunting task of applying for jobs. Chapter 8 tells the story of how digital ads have grown on our favorite internet sites like mold on a forgotten piece of cheese in the fridge, how it has become much harder to find good deals on Amazon and good info on Google, and what you can do to force these sites to show you what you're actually looking for. This helps you use Google to look for job openings without being flooded with scammy recruiter sites. Chapter 4 explores an important formula from probability theory, how professional gamblers have used it to beat lotteries and casinos, and how you can use it to help tame tasks that involve a degree of uncertainty. When facing the long odds of your job search, you apply these lessons to develop an application strategy that factors some of the luck out of the equation. And you use the odds-made-easy material in chapter 6, which teaches a probabilistic way of viewing the world that helps us think more clearly, to ballpark the likelihood of your applications making it to the interview stage. This lets you figure out how many positions to apply for to have a decent shot of landing one.

But then you remember the lessons from chapter 5, where the key concepts and formulas underlying risk, and what you can do to mitigate it, are discussed. Submitting more job applications increases your odds of success only to a limited extent, because of something called *covariance*. Best to diversify your efforts more effectively. Supplementing your job applications with some professional school applications would pass the covariance test and provide the buffer you seek against bad luck, so you decide to do that. But what degree programs, and at which schools, should you apply to?

You've heard chatter about how AI is going to disrupt many industries, killing some jobs while creating others. Are your long-term prospects better if you go to law school, medical school, or business school? What about a six-week coding boot camp? With AI disruptions on the horizon, you can't just look at past employment statistics. You need to look at what the experts think will happen in the coming years. Chapter 3 covers a simple technique for improving the accuracy of expert predictions, so you leverage this to anticipate which industries will be safest from automation.

Now, which college degree programs have the best chance of helping you break into the AI-proof industries you just identified? Your first instinct is to look at the rankings published by *U.S.*

News & World Report and *Times Higher Education.* But as we'll discuss in chapter 2, big-name rankings like these are rigid, impersonal, and frequently gamed. Instead, you can easily make individualized rankings that better reflect what matters most to you. This helps you find programs that are a good fit for you and that will put you on the professional path you seek.

It's a tough world out there, and math won't solve all your problems. But it can help.

2

Rethink Rankings

Late Monday afternoon on September 18, 2023, students, staff, and faculty at Brandeis received a distressing email from the president of the university. Earlier that day, *U.S. News & World Report* had released its annual Best Colleges rankings, and Brandeis had plummeted sixteen crucial spots, from number forty-four to number sixty. It wasn't just a matter of pride. A plunge of this magnitude, taking the school out of the top fifty, could impact recruitment, alumni donations, and even employment prospects for graduates.

The president's email reassured readers that Brandeis had received the same "raw score" as in the previous year, so the drop wasn't the university's fault. It was due instead to *U.S. News* implementing "significant changes in methodology this year that led to many dramatic shifts in the rankings." As President

Liebowitz went on to explain, the report had "removed or decreased the weighting of indicators that were favorable to private institutions like Brandeis." The timing made this news especially unwelcome. Brandeis was just one month away from celebrating its seventy-fifth anniversary with a large, festive gathering— which would now take place in the shadow of a ranking debacle. The email concluded with a resolute proclamation: "We maintain that no ranking can define what school is the best option for any individual student."

Another victim that day was the University of Chicago, which lost its top-ten status as it fell from number six to number twelve. Here, too, the blame was cast on methodological changes. An official response from the university declared, "We believe in and remain committed to academics and the fundamentals that have long defined the UChicago experience—such as our smaller class size and the educational level of instructors, considerations that were eliminated from this year's *U.S. News & World Report* ranking metrics."

College rankings are a zero-sum game, so when someone goes down, someone else must go up. California's Fresno State jumped a whopping sixty-four spots on that fateful Monday. The local newspaper, *The Fresno Bee*, cheered on this accomplishment with an article headlined "Go Bulldogs! Latest national ranking highlights Fresno State's academic performance." The article also praised the football team's twelve-game winning streak—a feat that no doubt involved using weights, but not the kind President Liebowitz blamed for the ranking shifts.

U.S. News said the 2023 shakeup resulted from the largest overhaul of the system underlying its annual Best Colleges list in the publication's forty-year history. Top-100 colleges typically jostle up or down around two or three spots, but in 2023 the movement was nearly four times greater. To understand what happened, how Fresno soared while Brandeis and Chicago slumped—even though the raw data reported by the schools changed very little—we must dig into *U.S. News'* mysterious and mercurial methodology.

The first step in *U.S. News'* process is choosing a collection of things to measure across all schools. (These are the "indicators" mentioned by Brandeis's president, Ronald Liebowitz, but I'll do what many others do and call them *factors*.) Next, *U.S. News* assigns a weight to each factor, indicating how strongly the magazine wants the factor to influence the rankings. The factor with the biggest weight, clocking in at 20 percent of the total weight both before and after the recent methodological change, is peer assessment. In *U.S. News'* words, this is "a measure of how a school is regarded by top administrators at other institutions." That's right: the biggest factor in the Best Colleges lists is not a measure of cold, hard, objective data. It's a survey of the subjective opinions of some higher-ups who aren't at the school in question. In other words, it's a popularity contest.

The fifth-highest-weighted factor, currently at 6 percent, is faculty salaries. It's hard not to see this as a measure of institutional wealth rather than academic quality. Yet *U.S. News* weights faculty salaries more heavily than factors that are probably more important to students—such as first-year retention rate (5 percent), borrower debt (5 percent), and student-faculty ratio (3 percent). Even measures we typically think of as indicating how competitive a school is or how strong its students are (such as standardized tests, worth 5 percent), are given lower weight than faculty salaries.

U.S. News' massive methodological overhaul in 2023 was simply a reweighting of the factors in its recipe. Alumni giving, which ostensibly shows how favorably alums feel about their alma mater but really measures how wealthy the school's alums are, was reduced to zero. The weight on graduation rates for students who received need-based Pell Grants was increased. These are both good changes, in my opinion, as they move the needle away from wealth and toward helping less affluent students. But by keeping peer assessment at the top, and faculty salaries near the top, *U.S. News* ensured that the rich and famous schools largely kept their clout.

The change that seems to have had the biggest impact is that the weight given to small class size went from 8 percent to 0. Wealthy schools can afford small class sizes, so this does help level the playing field a bit. But small classes aren't *just* a measure of a university's financial resources; many students really do prefer them. At only seventy-five years old, Brandeis is much less wealthy than many peer institutions that have had hundreds of years to grow their endowments. Harvard's endowment is $53 billion; Columbia's is $13 billion; even Dartmouth's $8 billion dwarfs Brandeis's comparatively meager $1 billion. Brandeis has managed to maintain small class sizes despite its relative "poverty," but this achievement is no longer reflected in the *U.S. News* rankings.

U.S. News described its choice to cut class size from the factors as a strategic decision to "place greater focus on outcomes measures." I have a hard time understanding how faculty salaries are more of an "outcomes measure" than class size. As we'll see later in this chapter, the decision to cut class size may have been strategic in a different sense: damage control for a massive scandal that rocked *U.S. News'* college rankings the previous year. I'll let you be the judge when we come to that.

Seeing the arbitrariness in choosing which factors to include and how much to weight them—and seeing the "dramatic shifts," as President Liebowitz called them, that result when these arbitrary choices are varied—might make you more sympathetic to his remark that "no ranking can define what school is the best option for any individual student." If so, good: you're making progress toward rethinking rankings. But I do believe that rankings can help students choose colleges. We just need to personalize the rankings by letting each person choose factors and weights that reflect their values. To do this, we must look under the numerical hood of rankings to see in more detail how they work—and how they can be refashioned to better suit our needs. The benefits reach far beyond higher education. Rankings, and the weighted sums they're built from, appear in settings ranging from the stock market and inflation to international travel and consumer credit scores. We'll learn the math to make sense of these and to personalize many of them, helping you compare options and make thoughtful life decisions.

The Math of Rankings

The first challenge with building a ranking system is that the different factors are typically measured on such wildly different scales that it doesn't work well to compare them directly. The difference between a huge class of 250 and a tiny class of ten looks insignificant compared with tuitions that might range from $15,000 to $80,000. To address this, *U.S. News* uses a popular statistical technique called *standardizing* the data. This converts all the numbers to a measure of how far above or below average they are. This process involves the concept of a standard deviation, so let's take a look at that first.

In the US, the average male height is five feet, ten inches. Heights, like many things in life, are distributed in a familiar bell-shaped curve. The average indicates where this curve is centered, while the standard deviation indicates how wide it is. Consequently, the number of standard deviations above or below the average is a useful measure of how rare something is. For American male heights, the standard deviation is about three inches. So someone who is six foot four is two standard deviations above average, meaning it's fairly rare to meet such a person—but not as rare as meeting someone who is three standard deviations above average, namely six foot seven.

To standardize a factor, we first compute the average of the factor across all the entities being ranked, then we record how many standard deviations above or below average each entity is. This is called a Z-score. For example, the average graduation rate across all colleges in the US is around 55 percent with a standard deviation of twenty-three percentage points, so a school with an 89.5 percent graduation rate gets a Z-score of 1.5 while a school with a 32 percent graduation rate gets a Z-score of –1. Converting all factors to Z-scores puts them on the same scale. Instead of comparing dollars, days, donations, and whatever else, we're just comparing standard deviations. A score of one for tuition means the same thing as a score of one for class size: one standard deviation above average.

After all the data has been converted to Z-scores, the next step in the ranking process is to combine the Z-scores across all the different factors to produce a single overall score for each entity. The most straightforward approach is to simply add up the Z-scores, but this puts all the factors on equal footing. It's more common to use a *weighted sum*, which allows some factors to influence the rankings more than others. With this approach, a weight is chosen for each factor, then when adding up the Z-scores, each is first multiplied by the factor's weight. For example, if a school has Z-scores of +2 and –1 and the first factor is weighted at 70 percent while the second factor is weighted at 30 percent, then the school's overall score is $0.7 \times 2 + 0.3 \times (-1) = 1.1$. These weighted sums of Z-scores boil each entity's data down to a single number so that they can be ranked from highest to lowest. This is the math behind most ranking systems, and it is how *U.S. News* decides which are the so-called best colleges. *Times Higher Education* in the UK has a similar university ranking system based on a weighted sum of Z-scores; it uses thirteen factors organized into five groups.

When Brandeis's president said the school fell sixteen spots despite reporting the same raw score as the previous year, he probably meant that the school's unweighted sum of Z-scores hadn't changed much—Brandeis was above and below average on the same measures in roughly the same amounts. But by changing the weights, *U.S. News* caused the rankings to shift considerably. As discussed above, dropping the weight on class size from 8 percent to 0 percent was a big blow for Brandeis, as it was for many other private schools that prioritize an intimate classroom experience.

If you're a student for whom small class size is important, then the 2023 reweighting made the *U.S. News* rankings less useful to you. The other weight adjustments impact different students differently as well. First-generation graduation rate (previously unused by *U.S. News*, now 2.5 percent) could be very important for some potential students and unimportant for others. Same goes for Pell graduation rate (formerly 0 percent, now 3 percent). Alumni giving (formerly 3 percent, now 0 percent) tends to correlate with wealth in potentially problematic ways, but some students might nonetheless value being at a school that inspires graduates to give back later in life.

It's frustrating that *U.S. News* gets to decide what's important and we don't have any say in the matter. Thankfully, math enables us to break free from *U.S. News*' one-size-fits-all ranking system. We can create personalized rankings that include whatever factors we want and weight them however we wish. Since not everyone is comfortable using spreadsheets and computing standard deviations, I'll show you an easy way to make flexible rankings by hand. You'll even be able to include quirky nonnumerical factors like how much you like a school's cafeteria food and how willing you are to support the football team. And you can use this math to rank anything, not just colleges.

Personalized Rankings

Creating your own ranking system is surprisingly simple. First, pick the things you want to compare. Next, choose the factors you'd like to include. Pretty much anything is allowed as a factor. You don't need actual data, and you certainly don't need Z-scores. For each factor, you just need to decide which of the things you're ranking is the best, which is the second best, and so on. Next, choose how much weight you'd like each factor to count for; it's common to use percentages that add up to 100 percent, but any numbers will work. There are no rules or magic recipes for this step; just as *U.S. News* chooses the weights however it wants, so can you. The final step, and the only one involving any math, is to compute the weighted sum of your rankings across all your factors.

Let's try this with college rankings. It's fine to start with a small list of schools you came up with the old-fashioned way, using things like word of mouth and intuition. Suppose you're trying to decide between Tufts, UCLA, and Georgetown. For the factors, let's go with how much you like the nearest city, how much you like the weather, student-faculty ratio, and how good the basketball program is. The first two factors are entirely subjective, but that's OK: if they matter to you, it's fine to include them in your personalized ranking.

All three schools are near large, vibrant cities, but let's say you like DC the best since you're interested in politics and international affairs. Boston is second. Los Angeles is third: it's OK, but not too convenient for a college student without a car. For weather, Los Angeles is the clear winner and Boston the loser (sorry, Boston fans, but you know it's true). The US Department of Education's College Navigator website has lots of useful data that helps with college comparisons. From that site we find that Tufts has the best student-faculty ratio at 10:1, Georgetown is a close second at 11:1, and UCLA comes in third at 18:1. Since Tufts and Georgetown are so close in this regard, I'm going to call it a tie for first between them and place UCLA in third. You're in charge of your personalized ranking, so you get to make executive decisions like this—if you feel that ranking them 1, 1, 3 more closely matches the data than ranking them 1, 2, 3, then go ahead and do so. Finally, UCLA and Georgetown both have storied basketball teams, but I'll give the win here to UCLA since it has eleven national championships to Georgetown's lone win in 1984. Tufts does well in its division, but compared with these two basketball powerhouses, we've got to place it third. Let's summarize all these rankings in a table:

	City	Weather	Student ratio	Basketball
Tufts	2	3	1	3
UCLA	3	1	3	1
Georgetown	1	2	1	2

Now that we have the schools ranked for each factor separately, we can choose weights to combine these rankings into a single overall ranking. First, let's try using equal weights across all four factors, meaning 25 percent each:

Tufts: 0.25 × 2 + 0.25 × 3 + 0.25 × 1 + 0.25 × 3 = 2.25

UCLA: 0.25 × 3 + 0.25 × 1 + 0.25 × 3 + 0.25 × 1 = 2

Georgetown: 0.25 × 1 + 0.25 × 2 + 0.25 × 1 + 0.25 × 2 = 1.5

Since lower scores are better, like in golf, we have Georgetown on top of our personalized ranking, followed by UCLA and then Tufts. Now let's try adjusting the weights by giving more weight to weather (35 percent) and basketball (35 percent) and less weight to city (15 percent) and student ratio (15 percent):

Tufts: 0.15 × 2 + 0.35 × 3 + 0.15 × 1 + 0.35 × 3 = 2.55

UCLA: 0.15 × 3 + 0.35 × 1 + 0.15 × 3 + 0.35 × 1 = 1.6

Georgetown: 0.15 × 1 + 0.35 × 2 + 0.15 × 1 + 0.35 × 2 = 1.7

This change is enough to propel UCLA to the top, since it put more weight on UCLA's strengths and less weight on its weaknesses. Tufts came out in third place both times, and it did so when I tried a few other sets of weights, too. My take is that based on these factors, if weather and basketball are really important to you, then UCLA is the place for you. If you're not sure how important these are to you compared with the other factors in consideration, then it's safe to place Tufts in third, but I wouldn't use these rankings to choose between UCLA and Georgetown. There's too much dependence on the choice of weights to reliably do so.

You might feel awkward choosing factors and weights whimsically like this, but it's no less scientific than what *U.S. News* has been doing for forty years. Seeing how the sausage is made, and making some sausage of your own, might make you less willing to rely on rankings at all, and that's perfectly OK. My advice is to view rankings as one data point among many when making decisions like which college to attend. And when you do make personalized rankings, try playing around with the weights to see how robust your winners are. If something tops your rankings for a range of weights, that should give you more confidence that it really is your top choice.

Feel free to experiment and have some fun with personalized rankings. Try ranking movies you've seen or restaurants you've eaten at—not just according to reviews by critics and the public (although those could be included as factors) but according to things that matter to you personally. If you're deciding where to move next, try ranking cities. Don't be afraid to get creative with the factors you use: it's OK to consider things like weather, music scene, restaurant scene, public transit, affordability, and job prospects. It's usually not too hard to rank cities for each of these factors separately, and weighted sums allow you to combine a bunch of separate rankings into a single holistic ranking.

Perhaps you've decided to add a new puppy to your home but

are having trouble figuring out which breed is best for you. Suppose you've narrowed the field down to three contenders: Saint Bernard (my personal favorite), golden retriever, and Yorkshire terrier. Let's say the criteria that matter to you are being safe around kids, protecting the house, and not needing too many walks. Here's my amateur attempt to rank these breeds on these factors:

	Safe with kids	Protects house	Endearingly lazy
Saint Bernard	3	1	1
Golden retriever	1	2	3
Yorkshire terrier	2	3	2

For those curious, Saint Bernards are extremely gentle and patient with kids—tolerating a remarkable amount of poking, climbing, and tail pulling—but 150-pound objects moving at high speeds do pose a threat to little ones. If we set the weights of our three factors to be equal, then the Saint Bernard comes out the winner, though all three are pretty close. If we put a little more weight on safety with kids, the golden retriever jumps into the lead.

When making decisions like this, it's tempting to look for websites that rank the best dog breeds, or best cities, or best whatever you're interested in. I encourage you to go a step further than naively trusting one-size-fits-all rankings. Look for websites that rank dogs, cities, and the like on a range of different factors, then decide how much each of the factors matters to you and try some weighted sums.

One-size-fits-all rankings usually aren't as useful as they would be if their priorities, reflected in the choice of factors and weights, more closely aligned with yours. But their problems run much deeper than this. In her bestselling book *Weapons of Math Destruction*, Cathy O'Neil argues that when different institutions focus their efforts on the same set of metrics, they come out looking kind of the same. Take higher education, for example. Instead of each school defining its own strengths and drawing students who appreciate them, universities know what factors they need to score well on, and they all do essentially the same things to boost their scores on these factors. Upending *U.S. News*' hegemony would restore some of the distinctness and individuality of yore. Personalized rankings are a key to doing this. And as you've just seen, the math needed to make them isn't hard at all.

Another problem O'Neil discusses is that the more influential a ranking is, the greater is the incentive to game it. Baylor paid the exam fee for students to retake the SAT *after* they were admitted, hoping to boost its Z-score on that factor. Bucknell and Claremont McKenna found a cheaper approach: they sent false data to *U.S. News* to inflate the SAT scores of their incoming students. Iona College (now Iona University) went further and was caught fudging nearly all the data it submitted. Some of the factors and weights *U.S. News* uses have changed since these scandals, but what hasn't changed is the pressure colleges feel to do whatever it takes to come out on top. And with the math of rankings in hand, we can now see how they are gamed.

Gaming the College Rankings

My first faculty position was at University of Georgia, and during my time there the university proudly announced an initiative to hire a dozen new math lecturers. Their primary role would be teaching Calc 1, with a goal of getting class sizes under twenty. This sounded laudable, as calculus is the academic equivalent of a root canal for many students and a cozier classroom experience could help. But I was left wondering: Why just Calc 1? Won't it be a shock for students to go from a Calc 1 class with nineteen students in the fall to a Calc 2 class with one hundred students in the spring? The answer the administration gave was that Calc 1 is an academic pinch point, with many students turning away from engineering majors and premed aspirations due to poor performance in the class. Yes, but there was more to the story.

Back then, *U.S. News* still counted class size as a factor in its college rankings, and it carried a hefty weight of 8 percent. To come up with a university's class size score, the magazine calculated a weighted sum of five sub-scores: the number of classes with enrollments under twenty, between twenty and twenty-nine, between thirty and thirty-nine, between forty and forty-nine, and over forty-nine. Universities got a substantial boost by having a lot of classes with fewer than twenty students. But not all classes were included in that tally. My university's actions struck me differently when I learned that Calc 1 was one of the classes *U.S. News* looked for in its under-twenty category, while Calc 2 was not. This was arguably a mild form of gaming the rankings, but the next story is a whole other can of worms.

When the *U.S. News* college rankings were launched forty years ago, Columbia was ranked eighteenth. Over the years it steadily climbed, reaching a pinnacle in the 2022 ranking, when it was listed as second in the nation—tied with Harvard and MIT and topped only by Princeton. The next year, Columbia dropped sixteen spots all the way back down to number eighteen. This was a single-year fall from grace as large as Brandeis's in the subsequent year, but for a school so close to the top to plummet so abruptly was even more staggering. And in the case of Columbia, it wasn't caused by reweightings within *U.S. News*. It was triggered by one of the university's own faculty members.

Michael Thaddeus is a highly regarded math professor at Columbia. I've known him for years because he and I happen to be in the same research community. I even used a result from his PhD thesis in my own dissertation—a result that, ironically, is all about adjusting weights, but not the kind *U.S. News* adjusted. As accomplished as he is in theoretical mathematics, Thaddeus doesn't normally do the kind of work that lands you in the newspaper, so I was surprised when I found his familiar face (albeit sporting an unfamiliar pandemic hairdo) staring at me from the pages of *The New York Times* on the morning of March 17, 2022. It turns out that in his spare time, Thaddeus had done some numerical detective work and uncovered a massive scandal. To boost its national ranking, Columbia had allegedly submitted data to *U.S. News* that was, in his words, "inaccurate, dubious, or highly misleading."

Thaddeus detailed these allegations in a methodical twenty-one-page analysis he posted on his webpage. He framed his report as a critique not just of Columbia but of the college ranking industry more broadly. Word quickly spread, generating curiosity among the public, interest among journalists, panic among university officials, and embarrassment at *U.S. News*. After looking into the matter, *U.S. News* changed Columbia's number-two ranking in 2022 to "unranked." The following year, with more scrutiny placed on Columbia's data, the university received its number-eighteen ranking.

I'll highlight just a few of Thaddeus's findings here. First up, instructional spending. Columbia reported $3.1 billion, the highest among the six thousand institutions of higher learning reporting to the government. That's more than Harvard, Yale, and Princeton combined. This staggering figure helped Columbia to score very well on the financial resources per student factor, then weighted heavily at 10 percent. But Thaddeus found Columbia's $3.1 billion to be the result of creative bookkeeping, in which dollars spent at the university hospital were counted as instructional spending on the shaky grounds that medical students are educated there.

Next, the percentage of full-time faculty with a terminal degree, meaning the highest degree in the profession. Columbia reported an impressive 100 percent here. But Thaddeus argued that this literally could not be possible, pointing to sixty-six faculty members who appear to counter Columbia's claim. Among these are distinguished scholars, artists, and even a Nobel Prize winner. These eminent faculty members surely add tremendous value to Columbia, but they do not have terminal degrees.

And now, class size. Columbia said 82.5 percent of its undergraduate classes had fewer than twenty students—an astounding figure that surpassed all other schools in the top one hundred. Thaddeus ran the numbers and found that the actual value Columbia should have reported was somewhere between 62.7 percent and 66.9 percent. It was hard for him to pin down the exact number. He had to painstakingly go through online course directories, some of which were available only through the Internet Archive, a service that saves older versions of websites. If it was this difficult for someone at Columbia to verify the reported class size figure, what chance did officials at *U.S. News* have of doing so? And if Columbia was fudging this figure, how many other schools were, too?

One year after this 2022 ranking that Thaddeus lambasted, *U.S. News* published the 2023 ranking that opened this chapter. That was the ranking with "dramatic shifts," in President Liebowitz's words, stemming from the biggest reweighting of factors in the forty-year history of Best Colleges. Five factors were eliminated entirely that year. One of these, as we discussed earlier, was class size, which dropped from 8 percent to 0. Another was percentage of full-time faculty with a terminal degree, which dropped from 3 percent to 0. That's right: two of the five factors that *U.S. News* dropped in the aftermath of the Columbia scandal were factors that featured prominently in that scandal.

Correlation isn't causation, as they say. But all this makes me wonder how much *U.S. News*' recent methodological overhaul was really about "outcomes measures," as the magazine claims. From where I sit, dropping these factors sure looks like a convenient albeit ham-fisted way to address the manner in which these factors were apparently gamed.

It's unfortunate that institutions feel compelled to game the rankings and find shady ways of doing so. Much worse, however, would be if the rankings were gamed by the organization publishing the rankings to ensure that the institutions contributing the most money come out on top. To that, the World Bank says, "Hold my beer."

A Ranking Scandal Engulfs the World Bank

Beginning in 2003, the World Bank published an annual report titled *Doing Business* that ranked nearly two hundred countries according to their business-friendly laws and regulations. It was eagerly awaited each year by investors, business leaders, and politicians. But after seventeen years, publication halted abruptly when some employees claimed there were "data irregularities" in the 2018 and 2020 rankings. The international law firm WilmerHale was brought in to investigate, resulting in a damning report that implicated two towering figures in the world of international finance: Jim Yong Kim, who served as president of the World Bank until 2019, and Kristalina Georgieva, who was the World Bank CEO from 2017 to 2019. To understand what WilmerHale found, we'll first need to dive into the math behind the *Doing Business* rankings.

The World Bank chose forty-one relevant factors and grouped them into ten topics, or "indicator sets." For instance, the "starting a business" indicator set was composed of four factors: the number of procedures required to start a business, the number of days needed, the cost (as a percent of per capita income), and the minimum capital needed (also as a percent of per capita income). To put all the factors on the same scale, the World Bank used a popular alternative to *U.S. News*' Z-score conversion: each factor was rescaled to make the scores range from zero to one. Next, a weighted sum was used to combine all the different factors. The weights here were specified by a simple mathematical recipe that ensured all ten groups of factors had equal weight: the weight on each of the four factors in the "starting a business" indicator set was $\frac{1}{4}$, the weight on each of the three factors in the "paying taxes" indicator set was $\frac{1}{3}$, and so on.

The backdrop to the 2018 *Doing Business* report was that, in the lawyerly words of WilmerHale, "Bank management was consumed with sensitive negotiations over its ongoing capital increase campaign." Translation: the Bank was trying to raise money. Then-president Kim and then-CEO Georgieva were overseeing the capital increase campaign, and Georgieva once stated that the Bank would be in "very deep trouble" if the campaign fell short. Where does the Bank raise money from? Its 189 member countries—the very same 189 member countries that it ranks in the annual *Doing Business* report. Perhaps you can now see why the Bank might have been tempted to rig the rankings. But if the weights were out of its hands, what could it do? That's the question WilmerHale sought to answer, and here's what it found.

Leading up to the 2018 report, government officials from China, a key shareholder of the Bank, complained to President Kim about China's disappointing number seventy-eight spot in the previous year's ranking. President Kim responded that "the report's methodology might require an update." When a draft of the *Doing Business* report came out and placed China eighty-fifth, Bank employees sprang into action. At one meeting, the *Doing Business* team was instructed to see what would happen if data from Taiwan and Hong Kong were included in China's data. They reported back the following day that this would raise China's rank to number seventy. One official approved this change, and the revised report was to be sent to the printer—but another official intervened to say that the issue was "still under discussion."

CEO Georgieva told the *Doing Business* team that Hong Kong's data could not be incorporated for political reasons. After some other failed attempts to boost China's ranking by adjusting the way the factors were computed, a senior official wrote to Georgieva: "In light of the fact that we have considered several alternatives to recasting the data, and found none of them satisfactory, I suggest we go back to Plan A, which is to present the actual *Doing Business* numbers." But the *Doing Business* team was pressured to continue looking for ways to improve China's ranking. They sought data points for which it might be "reasonable" to question the value and adjust if possible in a way that caused the "least damage" to other data in the report. In other words, they couldn't find an acceptable methodological change that would elevate China, so they searched for ways to doctor China's data that were as inconspicuous as possible.

The team found three data points they felt could be altered. For one, they reinterpreted a Chinese law on secured transactions to boost China's score in the factor measuring legal rights. One *Doing Business* leader told WilmerHale that "this data point was an ideal vehicle to accomplish the objective because . . . the law was unique to China, [so] the change would not cause ripple effects for the data of other countries." In the end, the Bank's creative bookkeeping managed to improve China's ranking from eighty-five to seventy-eight. This may seem like a lot of work for a small improvement, but that just shows how high the stakes were. WilmerHale documented similar shenanigans in the 2020 *Doing Business* report, though there the "irregularities" centered on Saudi Arabia, the United Arab Emirates, and Azerbaijan.

This debacle put an end to the World Bank's *Doing Business* report, at least for now. What about the two finance bigwigs implicated in the scandal? Jim Yong Kim stepped down as president of the World Bank on February 1, 2019, only a year and a half into the five-year term he was appointed to serve. Kristalina Georgieva took over as interim president, but she left soon after to lead the International Monetary Fund (IMF). Two years into her term there, the bombshell WilmerHale report came out. The IMF held eight meetings on the matter and determined that the evidence "did not conclusively demonstrate that [Georgieva] played an improper role" and that she would be kept on as the head of the IMF.

Numbers lend rankings a veneer of mathematical impunity, but scratch the surface and rankings reveal their dirty secrets. The choice of which factors to include, how much to weight them, and how exactly to measure them is tremendously important. But it's also highly subjective. Often, these important methodological details don't align with your values—and sometimes they conspicuously align with the financial incentives of the organization publishing the ranking. Meanwhile, the institutions being ranked feel pressure to report the most favorable data, rather than the most factual data. The more influence a ranking has, the more people on each side of the fence—those being ranked and those doing the ranking—are driven to game it.

Personalized rankings address all these unfortunate issues. By choosing the factors and weights yourself, you're more aware of the subjectiveness involved—and you can vary the weights to see how robust your rankings are. While colleges try to game *U.S. News* and countries try to game the World Bank, nobody will bother gaming the rankings you put together in your living room for yourself. And when you make personalized rankings, you can use data that isn't collected for the purpose of big national rankings and hence is less likely to be doctored. You can even use data that isn't data in the traditional sense—including the opinions of your friends and your past experiences.

To me, personalized rankings feel much more honest than

the big one-size-fits-all rankings. They don't mislead us with a false air of impartiality. They are simply tools to help people combine multiple considerations when making decisions. That's what all rankings are, but we tend to read far too much into the big official ones. I hope that by seeing all the chicanery around those, you'll treat them with skepticism and be more open to making your own rankings instead.

Not Just Rankings

Weighted sums are a popular ingredient in rankings, because they turn a collection of separate scores across a range of factors into a single aggregate score. And they do so in a flexible way that allows the factors to have varying amounts of influence. Weighted sums are used for this purpose in many settings, not just those involving rankings. The lessons you've learned so far—how rankings are made, how they can be used and abused, and how you can personalize them—directly carry over to these other applications.

Have you ever wondered what it means when they say on the news that inflation is 3 percent (or 5 percent, or . . .)? Surely this means that prices are going up 3 percent per year—but which prices? Housing prices, gas prices, food prices, consumer electronics prices: these are all important, but they often move at different rates and sometimes even in different directions. To get an overall sense of how much prices are changing, we'd need a measure that combines all these price factors. The weighted sum heeds the call of duty.

The Consumer Price Index (CPI), for example, is a weighted sum of prices across eighty thousand goods and services in the US. The weights are not chosen arbitrarily; they are meant to capture how important each item is to a typical American household. Americans currently spend about ten times more on meat products than on tofu products, so meat is weighted ten times more heavily than tofu in the CPI. The collection of goods and services in this weighted sum is fixed from year to year (except for infrequent periodic updates), so the percent the CPI goes up annually can be viewed as *the* inflation rate in America.

But what if you're an American who *doesn't* eat ten times more meat than tofu? Just as you can personalize college rankings, you could create a personalized consumer price index if you wished. No need to use eighty thousand items—just think of a handful of things you tend to spend money on each year, then compute a weighted sum of their inflation rates with weights that reflect what fraction of your budget each item occupies. As a simple example, if 30 percent of your paycheck goes to rent, 10 percent goes to food, and 4 percent goes to clothes, then you could compute

$$0.3 \times X + 0.1 \times Y + 0.04 \times Z$$

where X is the percent increase in your rent from last year, Y is the percent increase in your food expenses, and Z is the percent increase in clothes. The percentages here don't need to add up to 100 percent; it's fine to focus on just a few items in your budget rather than a full breakdown. If you don't pay rent, then don't include it in your personalized CPI, and if you buy a lot of clothes, then increase the weight on your clothes factor accordingly. If you get a cost of living adjustment to your salary each year, it's nice to know how it stacks up against the rise in prices for the things you buy, and the amounts of them you buy, rather than naively pretending that rising prices impact us all equally. Your neighbors don't spend money on the same things as you, so their inflation rates are not the same as yours.

Even when personalization isn't an option, it can be interesting just to recognize that a weighted sum is behind an important number. And it can be helpful to see the weights involved. Let's look at a few examples.

Did you know that credit scores are weighted sums? FICO is the most popular credit score in the US, and while the full details are a closely guarded corporate secret, FICO revealed that there are five main factors that are combined via a weighted sum. Payment history (including bankruptcies and late payments) carries 35 percent of the weight; amounts owed (including number of accounts and balances) has 30 percent; length of credit history (including the average age of accounts and the oldest account) has 15 percent; credit mix (having different types of credit, such as mortgage and consumer finance) has 10 percent; and new credit (opening new lines of credit and credit inquiries resulting from applications for new credit) has 10 percent. FICO won't tell us many details beyond this, but already this weighted-sum description helps dispel some of the mystery—and suggests what to focus on to improve your credit score. For instance, it's more important to avoid late payments than credit inquiries.

Outside the US, lending agencies tend to use their own formulas rather than relying on a centralized one like FICO. These formulas are almost always weighted sums; some lenders reveal the weights and factors while others do not.

If you follow financial news, you've likely come across various stock market indices around the world. Most of these are in essence weighted sums. The S&P 500 and Nasdaq in the US, the FTSE in the UK, the DAX in Germany, the CAC in France, the KOSPI in South Korea, the Hang Seng in Hong Kong, the Ibovespa in Brazil, and the NIFTY in India, to name a few, are known as capitalization-weighted indices. Here's what that means. The market capitalization of a company is the stock price of the company multiplied by the number of shares traded on the market. Capitalization-weighted indices are fractions where the numerator is the total market capitalization of the companies in the index and the denominator is a bookkeeping device meant to keep the index on a reasonable scale and to ensure that its value doesn't fluctuate too much when companies are added or removed. The numerator is what's most important, and one way to think about it is as the sum of the stock prices of the companies in the index weighted by the number of shares available for each. (The Dow Jones in the US and the Nikkei in Japan sum up stock prices without weighting them by the number of available shares. Some experts have argued that these indices are less meaningful because a big swing in a company's stock price has a big impact on the index, even if very few shares of the company are traded on the market.)

The travel journalists Asher and Lyric Fergusson publish a

popular "LGBTQ+ Travel Safety Index" that grades countries on how safe they are for members of the LGBTQ+ community to visit. The Fergussons note that "one cannot look only at data on whether or not same-sex marriage is legal and if anti-LGBTQ+ discrimination laws are in place. It also depends on the general attitude of the culture, minutiae of the legal system, and oppression of LGBTQ+ rights. These issues can affect everything, from your ability to show public displays of affection to being able to share a hotel room bed to the capacity at which you can use dating apps without being caught by the local police." To incorporate these different factors, they use—you guessed it—a weighted sum.

The Fergussons use seven factors with positive weights: legality of same-sex marriage, LGBTQ+ worker protections, legal protections against anti-LGBTQ+ discrimination, criminalization of hate-based violence, adoption recognition, transgender legal identity laws, and a Gallup poll asking locals if it's a good place for gay and lesbian people to live. They also include four factors with negative weights, meaning factors that lower the safety grade: transgender murder rates, illegality of LGBTQ+ relationships and acts, propaganda and morality laws, and illegality of changing gender. They score each country on these eleven factors, using scores that range from zero to fifty for all the factors except for the Gallup poll, transgender murder rates, and illegality of LGBTQ+ relationships and acts, which are scored from zero to one hundred. This means these three factors (one positive, two negative) are given twice the weight of the other factors, as the Fergussons believe they are more important for travel considerations. The seven countries receiving an A grade in 2023 were Spain, Portugal, Norway, Malta, Netherlands, Sweden, and Canada. At the bottom, seventy-two countries received an F grade.

Weighted sums are everywhere. You'll see in chapter 4 that they are behind one of the most important formulas in probability theory, and in chapter 7 that they shape what you see on social media. Now that you've spent some time with weighted sums, you'll be better equipped to navigate a world that relies on them to rank and score a wide variety of things. Most important, instead of trusting the big one-size-fits-all rankings that everyone tries to game, I hope you'll try making some personalized rankings. And when you can't avoid being scored by a weighted sum, I encourage you to look into the factors and weights behind the score so that you can play this numbers game as strategically and successfully as possible.

3

Make Predictions
Like a Pro

One of the more entertaining activities at the annual West of England Fat Stock and Poultry Exhibition in Plymouth, England, is a guessing game in which attendees predict how much an ox will weigh after it is slaughtered and dressed. One year, after the competition had concluded, a curious-minded man asked the event organizers if he could borrow the paper tickets on which the guesses were written. He put the eight hundred tickets in order, from lowest guess to highest, and then repeatedly tossed out the top and bottom tickets until he arrived at the middle guess. In statistical parlance, this is a kind of average known as the "median." (The average where you add values up then divide by the number of values you added is called the "mean.") When the curious man did this, he found something striking.

The median guess for the weight of the dressed meat was

1,207 pounds, while the true weight turned out to be 1,198 pounds. That's an error of only nine pounds—less than 1 percent. The man was stunned. He later wrote, "The result is, I think, more creditable to the trustworthiness of a democratic judgment than might have been expected." In other words, this idea of combining different opinions by averaging them, which he viewed as an example of collaborative reasoning, fared far better than he anticipated. Its success surprised him because he didn't know the mathematics of variance and covariance that underlie why it works so well—and that we'll explore in this chapter. The problem wasn't that he failed his introductory stats class in college; it was that the field of statistics as we know it today was still being born. The year was 1906 and the man was Francis Galton, a cousin of Charles Darwin and one of the founding fathers of statistics (and also one of the founding fathers of eugenics, but that unsavory subject is for another book).

Galton's analysis at the 1906 Plymouth exhibition, which he published the next year in the illustrious research journal *Nature*, was the birth of a powerful method for making predictions that is widely used today. To see a modern manifestation, let's look to one of the biggest names in the field of prediction. Nate Silver started out as a quantitative baseball analyst (think *Moneyball*). But at age thirty he became a national media sensation, and something of a data-crunching folk hero, when he pivoted to politics and successfully predicted the outcomes of forty-nine of fifty states in the 2008 US presidential election. He did even better in 2012, correctly predicting all fifty.

While Silver was a numerical wunderkind (in kindergarten he apparently enjoyed multiplying two-digit numbers in his head), the technique that drove his fantastic forecasts was something surprisingly simple: a weighted average. This is a flexible way to combine a bunch of predictions into a single prediction that usually outperforms each of the individual ones. The technique is similar to the one Galton used in Plymouth, but with a weighted average, you can place more emphasis on the predictions you have more confidence in. For Silver, these high-confidence predictions were the political polls that performed well in previous elections.

The idea of using weights to combine different numbers should sound familiar. The weighted *sums* we studied in chapter 2 help when you want to rank or score options that depend on multiple factors: what college to attend, what dog to adopt, how safe a country is to visit, and so on. The weighted *averages* that we'll study in this chapter are useful when there's an important number you want to estimate or predict: the price of a stock next year, the total snow accumulation tomorrow, the votes a political candidate will get in an upcoming election, the weight of a butchered ox. Before turning to weighted averages and how you can use them to make predictions that matter in your life, let's try something a bit odd: counting marbles.

Marble Madness

Imagine someone places a jar of marbles in front of you and asks you to guess how many there are. You could go with your gut and say whatever number pops into your head first, but what if you want a smarter strategy? Perhaps you could count the layers of marbles, then multiply that by the number of marbles in each layer. But it's tricky to figure out what even counts as a layer when the marbles are a jumbled mess—and since a jar is round, not rectangular, counting the marbles in each layer is no simple task.

There's something you can do that's both a lot easier and a lot more effective: ask a bunch of people to guess the number in the jar and then take the average of their guesses. It doesn't sound like that would do much, but it works remarkably well. The reason is that some people overestimate while others underestimate, so taking the average balances out the errors to arrive at a Goldilocks sweet spot.

There's more to the story, but to get there it helps to switch from one jar of marbles to several of them. The averaging technique works fine for a single jar, but the mathematical concepts that help us understand more deeply how and why it works become clearer when multiple guesses are involved. So now imagine a row of jars filled with marbles, and the task is to guess how many are in each.

The first thing we need to do is quantify how far off the mark each guess is. Let's use the word *error* here to mean the true number of marbles minus the number guessed; this will be a positive number for an underestimate and a negative number for an overestimate. And let's call someone a *balanced* guesser if the average of their errors is close to zero. This means they might guess too high on some jars and too low on others, but overall they're not systematically overestimating or underestimating.

At first you might think that being a good guesser is just a matter of being balanced. But there's an important subtlety lurking here. Suppose we have two jars, the first with 100 marbles and the second with 200. Someone who guesses 99 and 201 has perfectly balanced guesses, because their errors are one above and one below in each case, and the average of +1 and –1 is zero. Someone who guesses 98 and 202 has errors +2 and –2, which also average out to zero. Some maniac who guesses 0 and 300 is way off the mark, yet their errors are 100 and –100, which again average out to zero. Even this maniac is perfectly balanced! So if it's not just a matter of being balanced, how do we determine whether someone's a good guesser?

The key is a concept in statistics called *variance* that measures how wide a distribution is—or, in simple terms, how spread out a collection of numbers is. The average of one's errors indicates whether one tends to produce overestimated, underestimated, or balanced guesses; the variance of one's errors indicates how wild one's guesses are. The best guessers are balanced but not wild. Let's explore this with an example.

Someone who always guesses fifty less than the true value will have an average error of fifty, making them a systematic underestimator, while the variance of their errors will be zero. This person is not a good guesser, but they aren't wild at all. They're impeccably consistent, actually: they're wrong in the exact same way every time. Now imagine this robotically steady marble guesser has two friends. The first has a sequence of errors +5, –12, –7, +20, –6; the second has errors +35, +42, –67, +48, –58. Both have an average error of zero, but the second friend's errors have a much higher variance—and I think you'd agree that the second friend is a worse guesser than the first. To be a good guesser, you need to have an average error that's close to zero *and* a variance that's small.

We need just one more concept from statistics. The *covariance* of two sequences of numbers measures how similar their movements are. A large positive covariance means that when the numbers in the first sequence go up, those in the second sequence tend to go up as well, and when the first numbers go down, the second ones do, too. A large negative covariance means that they move in opposite directions (one sequence goes up when the other goes down). A covariance of zero doesn't exactly mean the two sequences are unrelated to each other, but that's a reasonable way to think about it in most real-world situations.

We now have all the mathematical preliminaries in place and can explain what makes averaging guesses work so well. To do so, let's suppose we have marble guessers Ariel, Benjamin, and Camila. And let's suppose Camila makes her guesses by averaging the guesses of Ariel and Benjamin. If Ariel and Benjamin are both balanced guessers, then Camila is guaranteed to be one as well. That's helpful to know, but it isn't too surprising. What's more astonishing is what happens when there's no covariance between Ariel's and Benjamin's errors, meaning one of them overestimating certain jars doesn't increase the odds that the other overestimates those same specific jars (and ditto for underestimating). You might expect that Camila's variance is simply the average of Ariel's and Benjamin's variances, but in this no-covariance situation her variance is *half* this average. That means she's twice as accurate as her friends, even though all she's doing is averaging their guesses! And it doesn't stop at two friends: the more friends she averages, the more her variance drops, and so the more accurate her guesses become.

Even if the covariance between Ariel and Benjamin isn't zero, averaging their guesses helps as long as their covariance isn't too big. This makes sense intuitively: if Ariel tends to overestimate the same jars that Benjamin overestimates, and they tend to underestimate the same jars as each other, then averaging their guesses isn't going to help Camila get closer to the target.

If Camila thinks that some of her friends are more reliable marble guessers than others, she can use a weighted average to prioritize the ones she has the most confidence in. The formula for a weighted average is easy. When adding up the numbers to be averaged, you first multiply each by its weight. Then, instead of dividing by the number of things being averaged you divide by the sum of the weights. If Ariel guesses 80 marbles and Benjamin guesses 120 marbles, then without weights Camila would guess $(80 + 120) / 2 = 100$. If Camila trusts Ariel 50 percent more than she trusts Benjamin, then she could give Ariel weight 3 and Benjamin weight 2, and the weighted average of their guesses would be $(3 \times 80 + 2 \times 120) / (3 + 2) = 96$. Notice that by putting more weight on the person who made the lower guess, the weighted average is pulled down a bit from the unweighted average of 100.

Weighted averages still manage to improve accuracy by reducing variance, but they provide an additional flexibility that unweighted averages don't offer. Let's see how you can use weighted averages in situations that are probably a little more important to you than the guessing game we played in this section.

Putting Weighted Averages to Use

When choosing the best time to refinance your mortgage, and whether to go with fixed rate or floating, you need to know not just what the rates are today but what they're expected to be in the future. When choosing how to invest your money, you'll want to know what prices different stocks are poised to reach and how much various industries will grow. From cable TV to financial newspapers and websites to investment apps, you'll likely find plenty of predictions for these quantities—and plenty of disagreement among the predictions. In these situations, weighted averages usually provide more accurate predictions.

Try combining traditional sources like *Bloomberg* and *Forbes* with something a little wilder like Jim Cramer on CNBC. Cramer's off-the-cuff style might be less accurate in general, but sometimes he sees something the others miss, so you could include his prediction in your average but give it a lower weight. Just make up numbers, it's OK. For instance, you could give a data-driven tool on *Bloomberg*'s website a weight of three, your favorite *Forbes* columnist a weight of two, and Jim Cramer a weight of one. Whatever weights you use, you're almost surely better off than not using averages at all and relying on a single source—thanks to the math we saw with marbles.

You can also use weighted averages to make decisions that

involve predictions of a more personal nature. Imagine you're in college trying to plan your career path and prepare accordingly. You want to do something creative that lets you lean into your artistic talent. Your school has a graphic design program that sounds promising, but before committing to it you want to find out how long after graduating it'll take you to land a job as a graphic designer and what you'll earn when you do. You can usually find these kinds of employment statistics on the website Glassdoor, but the numbers you see there are averages that don't reflect the details of your program or the strength of your portfolio. Ask some professors in the graphic design program at your school how long it took recent grads to land their jobs and what their starting salaries were. You can even ask how they think a student with your skill set will fare after graduating. When averaging all these predictions, you can decide whether you want to put more weight on Glassdoor, which has the most data, or on the professors at your institution. These professors know more about you and your school's recent grads than a national website does, but they might be biased and paint an overly positive portrait of the program they run.

Predictions aren't just about money. When you check next weekend's weather, you might see different forecasts on your phone's default weather app, Weather Underground, and Accu-Weather. If it's an extra important weekend—whether we're talking outdoor wedding or day on the golf course—don't take a chance trusting a single source: find a few different forecasts and average the high temperatures, average the low temperatures, average the chances of rain.

Politics, Polls, and Pundits

When Nate Silver shocked the political punditry by correctly predicting ninety-nine out of one hundred states across the 2008 and 2012 US presidential elections, the key ingredient in his election forecast system was a weighted average. He took the vote predictions from a handful of polling organizations and averaged them using weights representing how accurate their polls had been in past elections. It's worth reflecting on that for a moment. Silver did not choose the single poll with the best historical track record. He included all the polls he could find, even ones with less impressive track records, but he placed less weight on those.

Shouldn't including low-quality polls make your predictions worse? No. There's no guarantee the best polls in the past will remain the best in the future. And even the consistently lower-quality polls can uncover important patterns overlooked by other polls. An old-fashioned phone-based poll might have a better track record than a relatively new poll based on social media, but you're better off listening to what both polls are trying to tell you. Weighted averages let you do this, and Nate Silver showed the world how well they work.

Then came 2016. On the morning of the presidential election, Nate Silver gave Hillary Clinton a 71 percent chance of winning. As we all know, she didn't. Silver's political prediction blog was so immensely popular that some quickly blamed him for giving Clinton supporters a false confidence that ultimately doomed her by reducing voter turnout: If she was going to win anyway, why bother skipping work to stand in line on a Tuesday? In Silver's defense, while he did predict a Clinton victory, he predicted a slimmer margin than many other analysts predicted.

Averaging polls works well if some of them overestimate one candidate's chance of victory while others overestimate the odds of the other candidate. If something skews all polls toward one candidate, then a covariance emerges that eliminates the benefit the average is meant to provide. Some experts believe this may have been the case in 2016. A number of people who voted for Trump may not have admitted their intentions to pollsters before the election, due to Trump's, shall we say, unorthodox personality and campaign. If so, the polls would have misleadingly tipped more toward Clinton—as would have the weighted average based on the polls.

People still debate whether Nate Silver's 71 percent was a reasonable estimate and Trump's victory was just the luck of the draw. If we could redo that fateful election a handful of times, we'd be able to see whether Clinton ends up winning about two-thirds of them, thereby vindicating Silver, or whether Trump wins more than a third of the time, indicating a genuine popularity that went undetected in the polls. Alas, elections are not repeatable experiments—they happen only once—so we'll never know what to make of Trump's surprise victory and the predictions it defied.

How have poll-based predictions fared since 2016? Nate Silver correctly called forty-eight of fifty states in the 2020 presidential election, missing only Florida and North Carolina. That year *The Economist* offered its first-ever presidential election prediction, using a sophisticated system that in addition to weighting polls also leveraged "fundamentals" like GDP growth rate and the incumbent's approval rating. It got the same states correct and incorrect as Silver. RealClearPolitics offered predictions by directly averaging polls, without adding any of the additional bells and whistles that Silver and *The Economist* included. It, too, got all but two states correct, missing only Georgia and Florida.

This sounds pretty good, but as *Washington Post* columnist Perry Bacon Jr. noted, forty states were easy to predict for anyone closely following the 2020 election. While these predictions technically had a 48/50 = 96 percent accuracy rate, it's more like they were just predicting the ten remaining non-obvious results, and for those states their accuracy was 8/10 = 80 percent. Still good, but not nearly as impressive. Bacon Jr. cautioned, and rightly so in my view, that "polls are valuable, but not if they are used as the political version of Vegas odds for the Super Bowl." His reasoning is that due to the electoral college system in the US, close presidential elections in this country hinge on a small handful of swing states—and even though averaging polls reduces prediction errors, swing states tend to be so tight that this often isn't enough to predict them reliably.

While we'll never fully understand the confluence of factors that led to Trump's electoral victory in 2016, we're all painfully aware of an event that dominated the second half of his presidency: the COVID-19 pandemic.

Pandemic Predictions

After repeatedly dismissing the severity of COVID and defending his administration's response to the pandemic, President Trump released guidelines for "Opening Up America Again." In justifying these guidelines, which were announced on April 16, 2020, Trump and his officials cited the pandemic predictions of one particular research group. The researchers, based at the University of Washington in Seattle, developed their predictive system with somewhat modest ambitions in mind: they wanted to help hospitals anticipate the need for equipment like beds and ventilators. They didn't expect to catch the attention of the White House or the ire of the epidemiology community. But that's what happened with the April 16 announcement.

At the time, there was no consensus on which way the COVID numbers were trending or how high the death toll would peak. The Seattle forecast was an outlier; it was far rosier than nearly all others. Many outside the White House felt that the administration's choice to rely on it and ignore the rest was a deliberate— and deceptive—way to try to scientifically support Trump's decision to hastily reopen the country. But what if the Seattle group, and Trump, were right and the rest were wrong?

The standard method among epidemiologists to forecast the spread of an infectious disease is a two-step process. First, properties of the ailment are measured empirically to determine both how readily it passes between individuals and how long people remain contagious once infected. Second, predictions are produced either by feeding these parameters into a math formula or by programming them into computer simulations that allow researchers to mimic more complex environments and interactions (think *The Sims*, but a lot less fun). What the Seattle group did, however, was very different.

The Seattle researchers were not epidemiologists; they were data scientists. As such, they took all the virology out of the equation and viewed the predictive problem as one of finding patterns and trends in data. What they did, in essence, was use COVID infection and death data from cities outside the US to create curves showing how the pandemic might unfold within the US. For example, to see what would happen here if we reopened businesses but maintained social distancing, they found cities in Europe or Asia that had already done so, then extrapolated their infection and death rates to our population.

One issue with this data-driven method is that the predictions fluctuated much more wildly than the traditional epidemiological simulations. The epidemiologists needed only a few data points, such as how long you remain contagious and how many people you'll likely give the virus to while infectious. Reliable estimates for these were obtained fairly quickly, then didn't change much over time. For the Seattle group, on the other hand, whenever fresh data came in from around the world—a weekly, if not daily, occurrence—all curves would be recalculated from scratch and the predictions would change accordingly.

Updating estimates in real time is pragmatic in some sense, but it's hard to use a prediction that changes constantly. Well, it was hard for state governments that needed to decide what policies to implement. For the White House, this variability was an advantage. Whenever the forecast for infections and deaths dropped, the administration was able to point to the decrease as a sign that its policies and protocols were working, even if the decrease was due instead to an update from overseas data. And when the forecast went up, this provided a convenient excuse for Trump and his administration's lax pandemic response: we crafted our policies when the experts didn't think things would be as bad, they could argue, and it's not our fault that these experts have since revised their predictions.

Looking past the issue of how useful a rapidly changing forecast can be, the question before us—and debated heavily at the time—is the following: Was the Seattle group's approach the naive and flawed work of people lacking expertise in how viruses spread? Or was it a modernization of the old-school epidemiological ways, bringing the predictive power of the twenty-first-century data revolution to a field that refused to get with the times?

Shots were fired on both sides. The director of the Seattle team confidently and combatively ridiculed the epidemiological approach: "We believe in fitting models to data, and not making an assumption and then saying how my assumption would play out in a hypothetical world." The epidemiologists struck back: "The [Seattle] model is an odd duck in the pool of mathematical models. I fear the White House is looking for data that tells them a story they want to hear, and so they look to the model with the lowest projection of death," wrote an epidemiologist at the Yale School of Medicine. "That the [Seattle] model keeps changing is evidence of its lack of reliability as a predictive tool. That it is being used for policy decisions and its results interpreted wrongly is a travesty unfolding before our eyes," added an epidemiologist at the Fred Hutchinson Cancer Center.

In the end, there is no clear answer. Both sides managed some successes, some massive overestimates, and some massive underestimates. As the pandemic rolled on, however, one forecast rose above the fray and generally produced more accurate predictions than either of the competing factions. Any guesses how the team behind it pulled this off? They combined multiple expert predictions into a single prediction by using—drumroll, please—a weighted average.

This team was led by Nicholas Reich, a biostatistics professor at the University of Massachusetts Amherst. For several years his lab had worked closely with the CDC on influenza forecasting. In 2019, the CDC designated Reich's lab one of two nationwide Influenza Forecasting Centers of Excellence. When the pandemic arrested the world's attention the following year, it was only natural for Reich's lab to start forecasting COVID numbers. The team's approach was to average the predictions from several different forecasting groups, giving more weight to those that performed the best in the preceding twelve weeks. This is a lot like Nate Silver's system for forecasting elections, except that for COVID, the data rolled in much more frequently than once every four years.

What we learned from Ariel, Benjamin, and Camila applies not just to predictions but also to estimates and guesses of all kinds. It even sheds light on situations where the thing we're trying to figure out isn't a number, at least not ostensibly so.

Judges and Juries

In a sense, courtroom trials are more complex versions of the ox challenge Francis Galton witnessed in Plymouth. Instead of estimating the weight of a butchered animal, jurors must estimate the probability that the defendant is guilty of the crime for which they are accused. If this probability is high enough ("beyond a reasonable doubt," as we often hear on television), then a guilty verdict is handed down. Trials aren't normally framed in probabilistic terms like this, but that's what's going on, whether we like it or not.

Lawyers and judges are experts in the legal system, and they know the laws of the land better than ordinary citizens. But they're no better at estimating whether a defendant is guilty. All anyone can do is hear the evidence and decide if it's sufficiently convincing. It's risky to entrust a single citizen with the tremendous responsibility of deciding the fate of a defendant, so we rely on a jury of our peers. Court systems have varying rules, but in the US, criminal cases typically require jurors to be unanimous to convict, while civil cases rely on a majority vote. Either way, we gain important insight into juries by contemplating the math of predictions.

A jury is usually better than an individual at assessing a defendant's guilt because, as with marble counting, accuracy is improved when multiple estimates are combined. But the more covariance there is among the jurors, the less one benefits from this wisdom of the crowd. Put simply, when jurors all think alike, the conclusions they reach are not much better than those of individuals.

This has a dangerous consequence that is sadly familiar in many courtrooms. Since juries are selected randomly, their demographics tend to reflect those of society at large. But members of majority populations often mistakenly believe that members of various minority populations are more likely to be criminals. This means jurors are prone to a covariance in which they overestimate the guilt of those against whom society is prejudiced. A jury of racists is no less racist than an individual racist juror; a jury of homophobic jurors is no less homophobic than a single homophobic juror. Diversity is needed in a jury to prevent the covariances that lead to discriminatory verdicts.

Similar logic applies to panels of judges and justices. Let's consider the US Supreme Court. This court hears cases that have already been decided by lower courts; its task is to affirm or reverse those decisions. To put this task in the mathematical framework of this chapter, we can think of it as a matter of estimating the likelihood that the lower court ruled correctly. The nine Supreme Court justices often split into a liberal bloc and a conservative bloc. Generally speaking, there is a high degree of covariance within each bloc. This means that for many cases, the nine justices act in essence like a weighted average of two—a liberal justice and a conservative justice—where the weights are the sizes of the blocs. This results in a concentration of judicial power in the larger bloc, and a commensurate tendency for the larger bloc to rule along ideological lines. A bench with low co-variances that are spread out more uniformly across the justices would result in more independent interpretations of the law. Alas, judicial nominations are usually driven more by political considerations than mathematical ones.

Democracy and Voting

There is a close connection between prediction and voting. That's because a vote is in essence a prediction of which outcome will turn out better. This connection was not lost on Francis Galton. He opened his 1907 ox-weight-estimation article with "In these democratic days, any investigation into the trustwor-thiness and peculiarities of popular judgments is of interest." In the next paragraph, Galton asserted that "the average competi-tor was probably as well fitted for making a just estimate of the dressed weight of the ox, as an average voter is of judging the merits of most political issues on which he votes." What was most on his mind at the time, it seems, is how well democratic voting systems capture the will of the people and the extent to which majority voting selects the best political policies. This line of investigation hearkens back to an even earlier era.

The Marquis de Condorcet was an important figure in the initial years of the French Revolution. In addition to advocating for a liberal economy and a constitutional government, he espoused forward-thinking, egalitarian views. This included support for women's suffrage, denunciation of slavery, and opposition to the death penalty. He was the main author of an early constitutional document that was presented to the French National Convention in 1793 but did not receive a vote. Several months later a different constitution was put forth, by a different group, that was ratified. Condorcet openly criticized it, which led him to be branded a traitor. An arrest warrant was issued, and after successfully hiding for almost half a year, he was caught and imprisoned on March 27, 1794. He died two days later. To this day, historians debate whether his death was suicide or murder. What they don't debate is that in addition to his political activities, he was an esteemed and influential mathematician.

One of Condorcet's most famous mathematical accomplishments is a result he published in 1785, now known as Condorcet's jury theorem. It goes as follows. Suppose a group of individuals is to decide on a matter by majority vote, and suppose further that there is one correct choice and one incorrect one. In contemporary settings, this could be a jury voting on the guilt of a defendant, or a national referendum on whether the Earth is flat. The theorem states that if everyone votes independently and has greater than 50 percent chance of getting the choice right, then the more voters are involved, the higher the likelihood the majority vote is correct. But if each voter's odds are less than 50 percent, then adding more voters decreases the chance that the majority gets it right.

Some have used Condorcet's jury theorem to defend democracy, saying that it establishes the validity of majority rule by showing that the majority is correct more often than the individual. Others have taken a more pessimistic stance, arguing that voters are more likely to get things wrong than right, which would put us in the second category of the theorem's conclusion: poor choices at the individual level are amplified when large groups vote, making it all but certain that the majority will reach the wrong conclusion.

Each of these interpretations should be taken with a generous grain of salt. In real life, voters don't vote independently and in a purely probabilistic manner, and most of what we vote on are subjective decisions rather than factual matters with clear-cut right and wrong answers. And the theorem says little about the democratic decision we spend most of our energy on: which officials to elect. Nevertheless, Condorcet helped open the door to an important and ongoing effort to use math to better understand democracy and voting patterns. Francis Galton helped carry this work into the twentieth century, and Nate Silver caught the world's attention with it in the twenty-first.

The idea of using weights to place more emphasis on some voices rather than others is a further link between voting and prediction. In ancient Rome, rather than a "one person, one vote" democracy, votes were weighted by each person's wealth and tribal affiliation. In corporate governance today, publicly traded companies weight investor votes by the number of shares they hold. One of the most important examples of weighted voting in the US is the Electoral College: the fifty states vote for the president, and each state's vote is weighted in rough accordance with its population.

Advice from a Pro

Molly Hickman is a member of Samotsvety. No, I'm not referring to the 1970s Soviet rock band. Hickman is part of a remarkable forecasting collective named after the Russian word for
gemstone. The group formed in 2020 and quickly rose to dominate forecasting competitions. At one tournament, they scored
more than twice as high as the next-best team. This was described as "an absolutely obscene margin." Hickman joined
Samotsvety in 2021 and recently offered some advice for less
experienced forecasters: Don't put all your eggs in one predictive
basket. Instead, use a weighted average to combine different predictions, as we've been doing throughout this chapter. "Average
them together, give more weight to the ones that make more
sense to you." And trust your instincts: if the weight assigned to
a given prediction feels too high, "listen to that feeling." Hickman says that "good forecasting thrives on a delicate balance of
math, expertise, and . . . vibes." When I asked Hickman to elab-
orate on this point, she said that the more you practice making
predictions, the more you develop an intuition for how to choose
the weights, even if you can't explain the logic behind the choice.

Later in our conversation, I wondered if a person's track record
in past predictions—for instance, in forecasting competitions—
could be used to anticipate whose predictions for future events
might be most accurate. To my surprise, Hickman bluntly responded, "I'm a huge skeptic of track records." She explained
that competitors often rack up points by carefully choosing the

quantities they are predicting, and their success on those specific challenges is not guaranteed to carry over to others. This reminds me of the age-old trap of mistaking expertise in one realm with expertise in all realms. Hickman encourages people to look at the method behind a prediction to decide how much to trust it, and hence how much to weight it. But she clarified that while competition track records don't always separate the very good forecasters from the very best, they typically have no trouble separating the good from the bad. And for many real-world situations, that's all we need.

In the end, using weighted averages for predictions is a lot like using weighted sums for comparing options the way we did in chapter 2. In each case, you may feel like you're doing cold, hard, objective math when you apply the formula, but it's important to remember that your choice of weights in the formula is largely subjective. For predictions, weights can sometimes be chosen in a scientific manner by assigning more weight to the predictions that fared better in the past, like Nate Silver did with elections and Reich's lab did with the pandemic. But you can do that only when the predictions you're averaging have performance histories that you can access. Absent that, it's up to you to decide how much weight to give to each, and there's no rule book for doing so. That might make you apprehensive about trying weighted averages. Just remember: even when choosing the weights is a matter of intuition, you'll almost always get better predictions with weighted averages than you will if you just try to guess what will happen in the future.

4

What to Expect When You're Expecting Value

After cofounding the cryptocurrency exchange FTX in 2019, Sam Bankman-Fried quickly became the public face of crypto. His affable visage and iconically rebellious hair graced the covers of newspapers and magazines. *Forbes* drooled over him and the $20 billion fortune he amassed by age twenty-nine. Paid celebrity endorsements of FTX from Tom Brady, Steph Curry, Larry David, and others secured Bankman-Fried's spot in the pantheon of pop culture. But his house of crypto cards came crashing down abruptly when, at age thirty-one, he was convicted on seven criminal charges. A prosecutor described Bankman-Fried's work as "one of the biggest financial frauds in American history." His parents, both Stanford law professors, sat in the front row of the courtroom holding back tears as the jury announced its verdict.

There are many threads woven into this tale of hubris, hype, incompetence, and deceit. It's a story we'll likely be unraveling for many years. But as a mathematician, one angle (no pun intended) stands out to me: Bankman-Fried viewed the world in staunchly probabilistic terms and relied excessively on a math formula known as the expected value formula.

The author Michael Lewis, whose successes include *Moneyball*, *The Blind Side*, and *The Big Short*, recently focused his talents on Bankman-Fried, producing a biography that reached the number-one spot on the *New York Times* bestseller list. In it, Lewis writes that "every decision Sam made involved an expected value calculation." In this chapter I'd like to unpack this peculiar statement, explore the role expected value might have played in Bankman-Fried's spectacular rise and fall, and discuss ways you should and shouldn't use expected value in your own life. Along the way, you'll see how expected value has been used to beat casinos and the lottery, and what it has to do with taking the SATs.

Embracing What Einstein Wouldn't

After revolutionizing our understanding of the universe, Albert Einstein resisted a new development in physics that we now take for granted: the world runs on randomness. Quantum mechanics says that nearly everything is fundamentally probabilistic. Particles don't have locations; they have likelihoods of existing at different places. What we previously thought of as points in space are more like impressionistic smears of paint that convey the contours of location without pinning it down.

This destroys the long-held view of the universe as an intricate clockwork machine. We can calculate what *might* happen but not what *will* happen. This deeply bothered Einstein and led to one of his most well-known quips, loosely translated as "God does not play dice with the universe." He wasn't opposed to quantum mechanics, a field he helped usher into existence. He just held out hope that if we dug deeper, we could find the hidden gears underlying quantum mechanics—laws that would precisely describe the movements of everything and ultimately complement Einstein's largely deterministic view of physics.

A century later, the probabilistic interpretation of quantum mechanics shapes our modern world. We use it to power MRIs, lasers, solar cells, electron microscopes, and the ultra-precise atomic clocks needed for GPS. We've learned to accept randomness and harness it as best we can.

This lesson extends far beyond physics. From the movement of stock markets to the spread of pandemics, we now believe that many familiar phenomena involve a degree of randomness that forces us to speak in terms of odds rather than inevitability. This doesn't mean that we give up on trying to predict the outcome of uncertain events. It means that we embrace probabilities to navigate a world built on randomness.

Some people adopt this perspective more fully than others. Sam Bankman-Fried embraced it with religious fervor. Lewis tells of an incident in which Bankman-Fried messaged his personal assistant that there was a 60 percent chance he'd go to Texas the next day. Flummoxed, she replied, "What does that mean, a 60 percent chance? I can't book 60 percent of an airplane and 60 percent of a car, or 60 percent of a hotel room in Texas." Working for Bankman-Fried would have been Einstein's worst nightmare.

The Texas trip was not an isolated incident. Every discussion, every decision was probabilistic to Bankman-Fried. A *yes* wasn't a *yes*, a *no* wasn't a *no*. These words were instead indications of high and low probability. We have all faced the awkward situation of reneging on a prior commitment, but Bankman-Fried was on a whole other level. He would freely commit to things knowing there was a chance he'd back out. Not a chance something would come up preventing him from following through on the commitment—literally just a chance he wouldn't do it. Plans were ephemeral and excuses unnecessary.

Going hand in hand with Bankman-Fried's probabilistic way of life was a utilitarian, perhaps even Machiavellian, morality, an ends-justifies-the-means focus on outcomes. When he told his assistant there was a 60 percent chance he'd go to Texas, I think he meant there was something of value in Texas, but there was a chance something more valuable would arise elsewhere—and if it did, he'd have no qualms about canceling his trip at the last minute to seize the new opportunity. The 60 percent number he mentioned was, I believe, his ballpark estimate that nothing better than Texas would turn up before his flight.

For someone with Bankman-Fried's obsessive fixation on numbers, there's an issue with making decisions solely by choosing whichever option is most valuable: we don't know how valuable an option is before we choose it. That's because if God plays dice not just with the universe but with all our earthly experiences within it, then every option we consider necessarily involves some degree of randomness. Maybe that flight to Texas was to sign a contract worth $100 million, but there was a chance the contract negotiations would fall through, rendering the trip worthless. Is the trip then worth $100 million or nothing? In an almost quantum mechanical way, it's essentially both at once, much like Schrödinger's famous cat that is both living and dead until we observe it and thereby force it into one of the two states. But how does any of this help Bankman-Fried—or any of us, really—make decisions in life?

The way out of this quagmire of quantum uncertainty is to admit that, yes, we don't know for sure how valuable any of our options will be. But we can try to estimate each option's *expected value*, a notion from probability theory I'll turn to next. Indeed, Lewis writes that Bankman-Fried "calculated and recalculated the expected value of each commitment, right up until the moment he honored it or didn't."

The Expected Value Formula

Let's start with a simple example. Suppose there's a game where you flip a coin. If it's heads, you get three dollars; if it's tails, you get one dollar. How much should you be willing to pay to play this game (assuming your goal is money rather than fun)?

Since the coin comes up heads half the time and tails half the time, on average you'd earn two dollars each round. This means you should play as long as the cost is less than two dollars per round. At such prices you're not guaranteed to make money every round, but you *will* make money if you keep playing many rounds. For example, if it costs $1.50 to play, then half the time you'd lose fifty cents and half the time you'd gain $1.50. On average you'd then earn fifty cents per round, and if you played one hundred rounds, you'd expect to earn around fifty dollars.

In more complex settings it's harder to intuit these things, but that's where the expected value formula comes in. It tells you how much you should expect to get, on average, in any situation involving some randomness or uncertainty—assuming you know how much you'd get from each possible outcome and how likely each outcome is to occur. The formula says to multiply the probability of each outcome by the value of the outcome, then add up all possible outcomes.

Since you learned about weighted sums in chapter 2, it's worth noting that you can think of expected value as a weighted sum. In fact, you can do so in two ways. Expected value is the sum of the values weighted by the probabilities, and simultaneously it's the sum of the probabilities weighted by the values. While both interpretations are mathematically valid, psychologically my preference is for the first one: it says expected value is the total value you'd get from all the outcomes when each is downgraded based on how unlikely it is to occur.

Let's try this for our coin flip game. The probability of a heads is 50 percent and its value is three dollars, while the probability of a tails is also 50 percent but its value is one dollar. So, the expected value is $0.5 \times \$3.00 + 0.5 \times \$1.00 = \$1.50 + \$0.50 = \$2.00$. On average, you expect to win two dollars each round. This is why you should be willing to play the game if the cost per round is less than two dollars. Below that price, the game offers you a positive expected net earnings, which means that playing the game would be profitable for you, on average. For instance, if it costs $1.50 to play each round, then on average you'd expect to net $2.00 – $1.50 = $0.50 per round.

Remarkably, this one idea is enough to understand how the insurance industry works. Let's consider health insurance. An insurer uses your medical records to estimate the probability of the various ailments that could afflict you. By estimating how much each of these would cost to treat, the insurer can compute the expected value of medical claims you might file. This represents the amount your insurer expects you to cost them. Insurers try to engineer the expected payments they receive from you—in the form of premiums, co-pays, and other fees—to exceed the expected cost of insuring you. This isn't always possible. If you are older or have serious preexisting conditions, you might be an expected loss for the insurer, but you're still legally entitled to coverage. That's OK, because insurance companies don't need to make money from each client; they just need to turn a profit overall. Fees from younger, healthier clients subsidize the cost of insuring clients who are expected losses.

To see if they're on the right side of this equation, and to help set rates accordingly, insurance companies again turn to the expected value formula. They estimate how much they expect to bring in and pay out each year across their entire customer base. And they have one further trick up their sleeve. All the money paid in premiums and other fees doesn't sit idly while it waits to be paid back out in claims. This money is invested in assets like stocks and bonds so that it generates revenue until it is needed. How do insurance companies know how much money they should keep liquid for urgent needs instead of investing like this, and how do they estimate the amount of revenue this invested money will bring in? The expected value formula, that's how.

In short, insurance companies can afford a few bad coin tosses as long as they earn enough from all the good ones to net a profit. The expected value formula is the guiding principle for doing this. Lotteries are similar, mathematically speaking. The government knows it'll lose money on some lottery players, but it uses expected value to price the tickets and prizes so that more money comes in from those who don't win than goes out to those who do. At least, that's what's supposed to happen, and what usually does. Later in this chapter you'll see the spectacular consequences when the lottery commission gets the numbers wrong.

The same principles underlie the way casinos operate. They want gamblers to win sometimes, just not too much and not too often. Expected value is the casino's key to ensuring it's on the right side of that equation. Fortunately for casinos, the randomness involved in shuffling a deck of cards or pulling the lever on a slot machine is much simpler than the randomness involved in getting cancer or having a heart attack. Casinos can work out all the probabilities needed for the expected value formula with incredible precision.

Unfortunately for casinos, gamblers are constantly on the hunt for tricks to bend the odds in their favor, like card counting.

That's because to beat the casino you don't have to win every time, you just have to get your expected value into positive territory. If you can do that, then you can trust the odds and turn a profit by heeding Tolstoy's wise words in *War and Peace*: "The strongest of all warriors are these two—time and patience." Casinos are well aware of this. All the flashing lights and freely flowing alcohol are choreographed to push players off their game plans and encourage reckless spontaneity.

In the end, two main factors separate professional gamblers from amateurs: (1) the ability to methodically stick to a plan for a long time, resisting temptation and impulsivity, and (2) a gambling strategy that edges the odds enough to achieve a positive expected value. To see how this works, let's visit the roulette wheel.

Roulette

A standard roulette wheel displays the numbers one through thirty-six with alternating black and red backgrounds. Since this is a math book, I can't resist mentioning a fun piece of numerical lore: if you add up all the numbers on a roulette wheel, you'll find something curious that some say was a deliberate design choice to add a demonic mystique to the wheel. Any guesses? At the end of this section, I'll show you a cool trick for adding up numbers arranged like this and reveal the answer.

When a croupier spins the roulette wheel, a little ball bounces around erratically, eventually settling into one of the colored numbers. Betting on red has a 50 percent chance of winning and a win earns what you wagered, whereas a loss counts as negative earnings because you forfeit your wager. Same for black. Betting on a single number has only a one-in-thirty-six chance of winning, but the lucky player who pulls this off earns thirty-five dollars for every dollar they bet. There are other types of bets you can place, and they all share a simple property with the bets described so far: the payoffs are calculated to have expected value zero. For example, if you bet ten dollars on a single number, then the expected value formula gives $(1 / 36) \times \$350 + (35 / 36) \times (-\$10) = 0$. This means you may have streaks of good luck and of bad luck, but on average you're going to break even if you keep playing roulette, no matter how you bet.

Except, what I wrote in the previous paragraph isn't quite true. In addition to the red and black numbers, there's also a green zero (and on some wheels, a green double zero). These green zeros don't look like much on a wheel with thirty-six red and black numbers, but they're enough to make the game profitable for the casino by nudging the expected value of all your bets into negative territory. Play long enough and you're all but guaranteed to lose money to the roulette wheel, thanks to these little green devils.

People have long wondered whether it's possible to use physics to calculate the trajectory of a roulette ball and predict where it will land. That sounds difficult, but we can land rockets on Mars and smash particles into each other at near the speed of light—is a bouncing ball in a spinning roulette wheel harder than impressive feats like these? The eminent physicist Stephen Hawking seems to have thought so. He once said of roulette, "It is practically impossible to predict the number that will come up; otherwise, physicists would make a fortune at casinos."

However, just as Einstein's reluctance to accept the probabilistic nature of quantum mechanics was misplaced, Hawking overlooked something crucial about the probabilistic nature of gambling. You don't have to predict the number that will come up in roulette. You just need to predict two numbers that *won't* come up. Why? Expected value.

The green zero is the casino's clever trick for shifting the expected value of the roulette player's earnings from zero to negative territory. If you can reliably eliminate one number each time the wheel spins, then your expected value bends back to neutral ground and you'll break even on average. If you can eliminate two numbers each time, then you've not only undone the advantage the casino gave itself with the green zero; you've taken that advantage for yourself and secured a positive expected value. Predicting a couple of spots the ball won't land on rather than the precise spot it will sounds much more manageable. But can it be done?

In the early morning hours of Tuesday, March 16, 2004, a trio of gamblers at London's high-end Ritz Club casino cashed out their earnings from a long and late night at the roulette wheel. One had turned £30,000 worth of chips into £310,000. Another went from £60,000 to £684,000. By the end of two weeks, the trio had earned £1.3 million from playing roulette.

The Ritz had seen bigger winnings at roulette, but this trio's style was unusual—and unusually consistent. They wouldn't place their bets right away. Instead, they'd stoically wait six or seven seconds after the ball was launched, then all leap up to bet with impeccable synchronicity "as if someone had fired a starting gun," in the words of one Ritz employee. They didn't bet on single numbers, or on colors (which alternate throughout the wheel). Their preference was to place "neighbors" bets, where you cover one number along with the two numbers on either side of it, giving a span of five adjacent winning spots.

Remember that the bets in roulette are designed to have expected earnings zero were it not for the green zero, which drops all bets to a small statistical loss for the player. A neighbors bet fares no better or worse than other types of bets in this regard. But these expected value calculations assume that all numbers on the wheel are equally likely to win. If you were able to determine that some regions of the wheel are more likely than others, then a neighbors bet could be a way to take advantage of this information and bump your expected value back onto positive ground.

Indeed, betting on a color or an individual number would require the kind of predictive precision Hawking ruled out. But for a neighbors bet to be profitable, on average, you just need to find a five-number span on the wheel that's a little more probable than others. Perhaps waiting a half-dozen seconds into the ball's fateful journey held the key to doing this.

After the conspicuous gambling trio left, the Ritz's staff scrutinized the videos and eyewitness testimony for any signs of cheating. *The Mirror*, a UK newspaper, claimed that these men were a high-tech gang using hidden lasers to measure the ball's movements and software that predicts the ball's destination. Casinos have uncovered various attempts to sneak in electronic devices to assist gamblers, but in this case a full investigation turned up no evidence of the sort. One of the managers at the Ritz said the presumed leader of this trio, a Croatian man named Niko Tosa, was the most successful player he's seen in his twenty-five years on the job. The question remained how Tosa pulled off this unparalleled success.

A reporter for *Bloomberg* took on the mission of answering this question. After some impressive detective work, he managed to track down Tosa, who proved a shadowy figure. Tosa was adamant that he never cheated by using any kind of electronic device. He said he had a roulette wheel at home and simply trained himself to anticipate what region the ball would land on through tedious, methodical practice. But should we take him at his word?

The *Bloomberg* reporter failed to resolve the mystery of Tosa's prowess, though he found gambling experts who said that some roulette prediction is possible without technology, in the right circumstances. If the wheel has any imperfections, even extremely minor ones that aren't visibly apparent, the distribution of the ball's landing locations can deviate from uniform, meaning some numbers will be ever so slightly more probable than others. And some people do have a knack for anticipating which region of the wheel the ball will land in—a knack that improves with practice and that benefits from the gambler waiting as long as possible in the wheel's spin before placing the bet, just as Tosa and his crew had done.

Before heading to Vegas to try this yourself, be warned that in the two decades since Tosa dominated the roulette wheel at the Ritz, some things have changed. To limit the advantage card counting provides, casinos today shuffle together more decks than they used to. Beating the roulette wheel is also harder than it once was—and it was never easy. Most casinos are more diligent about replacing old wheels with new, unblemished ones. Some even modified the wheel's design to add more chaos to the ball's bouncing path. And some shortened the period in which gamblers are allowed to place bets after the ball is launched.

But now you know the game: Professional gamblers do whatever they can to get their expected value above zero. If they can do this, then they trust the odds and play for hours on end, letting probability do its thing. It's not glamorous, but it is effective.

Remember the question of what the numbers on a roulette wheel add up to? Here's the clever trick for answering this. Imagine the numbers one through thirty-six arranged in a line, then copy that line but reverse the order:

1	2	3	. . .	36
36	35	34	. . .	1

Each column adds up to thirty-seven, so if you add the numbers in all thirty-six columns you get $36 \times 37 = 1{,}332$. But this count adds up all the numbers from one to thirty-six twice, so we need to divide by two to get the sum we seek: $1{,}332 / 2 = 666$. In Christian mythology this is known as the number of the beast, or the devil's number. The roulette wheel certainly has a long history of bedeviling those who think they can outsmart it.

Can You Beat the Lottery?

Some say that lotteries are a tax levied on those ignorant of statistics. I used to think so, too, until one summer afternoon in Manhattan, Kansas (the "Little Apple," as it's affectionately called). My uncle was a statistics professor at Kansas State University, and I vividly remember a scene from one of my childhood visits. Uncle Paul tried to sneak out the front door without us noticing, unsuccessfully. My parents asked where he was going. To pick up his weekly lottery tickets, he answered. But don't worry, he reassured us, he'll get us tickets, too.

My parents, who think of themselves as rational types, were incredulous, bordering on incensed. Lotteries are a waste of money, and of all people, a statistician should know this, they exclaimed. (They didn't go as far as saying the expected value is negative, but that's essentially what they were thinking.) They pleaded with him not to waste his money, on us or on himself, but to no avail. Off he went, as my parents continued to chastise him.

The afternoon wore on, but the skepticism and negativity slowly morphed into curiosity and fantasy. If we did win—which, of course, we wouldn't—what would we spend the money on? Plans to help family members were discussed, as were home renovations and some travel. But travel to where? Details needed to be debated at length, to properly prepare ourselves for the winnings that we certainly wouldn't see.

As our excitement built, so did our anxiety about the drawing. "Turn the TV on, we don't want to miss it," my dad commanded. "And make sure it's the right channel," my mom added. In their impassioned state, my ultra-rational parents seemed to have forgotten that Uncle Paul bought tickets every week and was in no danger of missing the drawing.

At last, it was time. The numbers were drawn. Sure enough, we didn't win a damn thing. "Ha, we told you it was a waste of money," my parents smugly scolded. "Nonsense," my sagacious uncle replied. "Where else could you buy two hours of dreams for a few bucks?" He was right. A lottery ticket isn't a calculated bet like Niko Tosa's at the roulette wheel. It's a ticket to temporarily experience the fantasy of financial freedom untethered from the oppressive shackles of reality.

I took from this experience a lasting lesson that expected value doesn't capture the full depth of life. It conveys value only in a limited, quantitative sense. Governments run lotteries because they make money. They make money because each player's expected value is negative. This means that, on average, players lose more money than they earn. But as my uncle showed us, this does not mean that playing the lottery is a waste of money. Many things in life have value beyond the narrow-minded kind manifest in a standard expected value calculation.

That said, on some rare occasions lottery jackpots get so large that the expected value, even in the strict financial sense, becomes positive. However, there's a problem for those wishing to take advantage and cash in when this happens. Positive expected value means the average winnings are greater than the average cost of a ticket. But lottery tickets aren't like Schrödinger's half-dead cat or Bankman-Fried's 60 percent trip to Texas. A ticket either wins or loses. So how do you partake in these mythical "average winnings"?

Mathematicians often like to understand things by considering extreme cases, and that's helpful here. No matter how big the jackpot is, if you just buy a single ticket, then the most likely outcome is that you lose the full price of your ticket and win nothing in return. But what if instead of buying one ticket, you bought *all* the tickets? For a typical lottery this would be a guaranteed loss of money, even though you'd be guaranteed to hit the jackpot. That's the basic pricing that leads to the usual negative expected value for players and positive expected value for the government. However, when the jackpot becomes so large that the expected value flips around, buying every ticket would be a guaranteed way to make money.

Ordinarily you can't buy all the lottery tickets—but you could buy a lot of them. When a massive jackpot pushes the expected value into positive territory, the more tickets you buy, the more the law of averages kicks in and increases the odds that your investment will be profitable. No matter the jackpot size, the more tickets you buy, the more your per-ticket returns approach the average earnings computed by the expected value formula. Put another way, the more tickets you buy, the less luck factors into the equation—and the expected value formula tells you whether a luckless lottery is one you should play.

In practice, there's one irksome challenge to basing a lottery strategy on the expected value formula. For most lotteries, your share of the jackpot depends on how many other winners there are. This means you can't just do a simple expected value calculation based on the price of a ticket, the size of the jackpot, and the odds of winning the jackpot. You have to incorporate the odds that someone else wins the jackpot, and that's harder to estimate since it depends on how many people decide to play. Nonetheless, people have at times successfully gamed the lottery using the ideas discussed here.

On April 25, 2023, headlines across Texas announced that a winning ticket for the $95 million jackpot was sold in Colleyville, a Dallas suburb. It's easy to imagine the lucky individual walking into a 7-Eleven and picking up some Doritos and cigarettes and a weekly lottery ticket, oblivious to how life-changing this purchase would prove. But that's not what happened. The winning ticket didn't come from a convenience store, and there was no "lucky individual" involved. This was the work of a strategic cartel expertly playing the odds.

Three days before the drawing, eleven million lottery tickets—including the winning one—were purchased from an inconspicuous fishing store in a Colleyville strip mall. This staggering volume made the little store the biggest lottery retailer in all of Texas, by a considerable margin. But the fishing store doesn't actually sell tickets, at least not in the traditional way you might expect. It is part of a Montana-based retail chain owned by a man named Kevin Kramer. This Kevin Kramer also runs a private company that sells lottery tickets online, but he can only legally sell tickets for the Texas lottery if he has a physical location in the state. The Colleyville fishing store conveniently serves this purpose.

Kramer told an investigator that his online store is known as one of a handful able to handle high-volume sales. This attracts the attention of gambling syndicates known in the lottery world as purchasing groups. The executive director of the Texas Lottery said that the $95 million jackpot "generated significant interest and participation by purchasing groups to buy large quantities," and that "it appears the winning ticket was likely bought by one of the purchasing groups." Ticket sales for the April 2023 drawing totaled $28 million, $26 million of which came from purchasing groups.

No laws appear to have been broken here. By pooling their resources and using high-volume online sales, purchasing groups simply take advantage of the positive expected value offered by massive jackpots like this one.

Why Play If You'll Lose?

Insurance companies, casinos, and lottery commissions all make money by engineering their payouts to have negative expected value for the customer. Most of the time they succeed and turn a profit. Stories of gambling syndicates beating state lotteries and roulette wheels are the exception, not the norm. But if insurers, casinos, and lotteries are statistically losing propositions for ordinary people, why do so many of us partake in them? The quick answer is that expected value is a type of average, and averages aren't everything.

Let's start with lotteries. I've already mentioned that expected value calculations overlook things of intangible value, such as an afternoon spent blissfully dreaming of a life transformed by a jackpot. I also mentioned that expected value tells us the value of an "average" lottery ticket, which is a rather fanciful concept. Each ticket is either a winner or a loser, not a quantum mechanical combination. This means that regardless of any expected value calculation, each ticket carries the potential to be a winner. That potential, however improbable, is the allure of the lottery.

Another issue with expected value is that, as with other forms of averages, it is heavily influenced by outliers. In the case of the lottery, this means the jackpot. Most people who play the lottery enjoy the smaller wins they get from time to time. But these impact the expected value formula far less than the big jackpot that you're almost certain to never see in your life. It's a bit like calculating the average salary of a company to get a sense of what it would be like working there, not realizing the CEO's multimillion-dollar income has no bearing on the wages of the company's factory workers.

When deciding whether to play the lottery—or to partake in any other form of gambling—you should look past the averages and delve into your individual circumstances. Ask yourself questions such as the following: Would you enjoy playing even if you lose? How else might you spend this money? If you saved this money instead of gambling it, how much impact would this have on your savings in the course of, say, ten years? Expected value largely ignores these more personal and subjective matters. Numbers are part of the story, to be sure, but they are seldom the whole story. This is especially true of averages, including expected values, since they blur away so many important details.

The shortcomings of a myopic fixation on expected value are even more apparent in the case of insurance. Expected value says that most people should roll the dice and take the chance of going without any insurance, since on average they'll come out ahead by doing so. But you are not an average; you are an individual. If you require an expensive cancer treatment, or your house burns down, or your car is totaled in an accident, averages and odds are of little solace. Insurance is a fee we pay to avoid the devastating consequences of catastrophic events.

Don't Make the Mistakes Sam Made

Bankman-Fried's meteoric ascent to fame and fortune was propelled in part by his zealous adherence to probabilistic thinking, particularly expected value, in all facets of his life. He played the odds and took chances when others chose more cautious routes. He let numbers guide the way when others relied more on intuition. He treated social interactions as calculated transactions, always attempting to maximize his own benefit despite the consequences. But this same pattern of behavior paved the way for his downfall. A detailed analysis of Bankman-Fried's rise and fall is beyond the scope of this book. But I would like to highlight a few details to draw two important lessons on how you can get the benefit of an expected value mindset while avoiding the pitfalls that sank Sam.

After leaving his job at the large trading firm Jane Street Capital, Bankman-Fried moved to Berkeley in 2017 and cofounded his own trading firm, Alameda Research. He astutely saw lots of money to be made in the burgeoning cryptocurrency market. And he knew that it was mostly a numbers game, so he directed his firm to act accordingly. Rather than investing in crypto companies and products building genuine value, Alameda Research shaved off profits from inefficiencies in the unregulated and highly speculative international crypto market.

This was a perfect setting to live and die by the probabilistic sword and worship at the altar of expected value. Alameda Research was in essence a professional gambling syndicate, but one focused on cryptocurrency rather than casinos. And as we've seen with the roulette wheel and the Texas lottery, gambling syndicates can do very well by leveraging the expected value formula. But Bankman-Fried took his expected value obsession far beyond his firm's trading strategies.

The first lesson to draw from the Bankman-Fried saga is that expected value calculations best capture direct effects and tend to fall apart when a complex web of downstream consequences is involved. That's because to compute expected value, you need to first list all the possible outcomes and then estimate how likely and how valuable each is. In many real-world situations, the further you look into the future, the more the range of potential outcomes splinters into a vast multitude that is hard enough to list, let alone assign probabilities and values to. When this happens, expected value loses its efficacy and, even worse, can seduce you into dangerously shortsighted choices.

Lewis's biography contains an entertaining scene in which the media mogul and fashion icon Anna Wintour video-called Bankman-Fried to see if he would attend, and perhaps also fund, the year's Met Gala. Lewis details Bankman-Fried's eccentrically calculated deliberations in which he tried to run a cost-benefit analysis of the situation. The question in Bankman-Fried's mind was whether sponsorship of the Gala would generate more revenue than it cost. His reasoning was that crypto had a heavy gender imbalance, so as expensive as the Gala was, if his association with the fashion event drove enough women into crypto, it might be a net win for him. Needless to say, he took a probabilistic approach to the analysis.

In the end, Bankman-Fried didn't give a firm response on the matter of sponsorship, but he did say that he would attend the Gala with Anna Wintour. Then, in a typically Bankman-Friedesque manner, he bailed shortly before the event when he reran the numbers and decided it wasn't worth his time. Wintour's people were livid. Lewis describes how the Met Gala debacle was just the tip of the iceberg: "Sam's last-minute decision not to go would not create anything like the havoc caused by some of his other internal calculations. CEOs had flown to the Bahamas under the mistaken impression that Sam had agreed to buy their companies. The World Economic Forum had to scramble to fill a stage and cancel media interviews after Sam decided, the night before he was meant to deliver a big speech in Davos, not to."

What does this have to do with expected value? Bankman-Fried could estimate the probability that his attendance at the Gala would bring a thousand more women into crypto and the value this would provide his investments. But what is the cost of breaking a promise to Anna Wintour? This is nearly impossible to quantify, so Bankman-Fried must have either guessed an ill-fitting number for it or simply ignored it entirely. Even worse, Bankman-Fried's expected value framework largely treats each decision in isolation, overlooking the fact that our actions are cumulative and have broad-reaching consequences. The cost of deceitfulness compounds when it is habitual, yet Bankman-Fried didn't seem to take this into account.

When his crypto empire began to crumble, Bankman-Fried exchanged text messages with a journalist that did not exactly paint a portrait of sincerity. When asked if his repeated calls for regulation of the crypto industry were just PR, he answered, "yeah just PR. fuck regulators. they make everything worse." When asked about an earlier statement he made about how people shouldn't do bad things even for the greater good, he replied, "man all the dumb shit I said. it's not true, not really." I can't imagine that any of this sat well with the jurors who would decide his fate one year later. They found him guilty of all charges: fraud, conspiracy, and money laundering.

This was a man who failed to recognize the obvious price of repeatedly breaking promises, bending the truth, and outright lying: people stop trusting you. What's less clear is whether this was a consequence of the flawed utilitarian expected value framework he espoused, or if it was simply who he was all along. Perhaps he was drawn to the expected value formula in the first place because it enabled and even encouraged the kind of amoral and shortsighted behavior he was naturally prone to. Either way, Bankman-Fried's lack of consideration for long-term consequences leads to the next lesson.

Expected value provides a reliable gambling strategy only when you can keep playing. A positive expected value says that on average you'll make money, so while you may suffer some losses, if you push through them, you'll eventually come out ahead. But if you lose all your chips, you can't push through—it's simply game over and the expected earnings you sought are but a mirage. Put another way, if a gambler repeatedly goes all in, eventually they'll go bankrupt no matter how good the odds are. This is one form of a principle in statistics known as *gambler's ruin*.

That's in essence what happened to Bankman-Fried, and in two ways. His trading firm Alameda Research gambled so aggressively that at one point it needed several billion dollars to cover its liabilities. This was the first strike of gambler's ruin. Rather than calling it quits and letting Alameda Research go bankrupt, Bankman-Fried "borrowed" billions of dollars from customers at his cryptocurrency exchange FTX, without their consent, to cover Alameda's losses. This was a legal gamble more than a financial one, and it did not fare any better—as his conviction shows. That's the second strike of gambler's ruin: if you break the law often enough and in serious enough ways, eventually you'll get caught. And it's not easy to rebuild a multibillion-dollar fortune from the inside of a prison cell.

Expected value works best when you can keep rolling the dice, but this dual dose of gambler's ruin spelled the end of Sam's game. You can't win if you can't play again.

So How Can We Use Expected Value?

We shouldn't let this cautionary tale of Sam Bankman-Fried scare us away from the expected value formula. It is used all the time to great effect by insurers, financial investors, public health officials, sports analytics practitioners, and many more. Legendary investor Warren Buffett once said, "Take the probability of loss times the amount of possible loss from the probability of gain times the amount of possible gain. That is what we're trying to do. It's imperfect but that's what it is all about." What he's describing is none other than the expected value formula—probability times value summed over the different possible outcomes. His point is that we don't know how any particular investment will turn out, but the expected value formula suggests when an investment is worthwhile.

You can use the formula beneficially in your own life, too—you just need to be mindful of the lessons Bankman-Fried inadvertently taught us about stretching the formula too far. Expected value is most practical in settings where there are only a small number of possible outcomes, where each has a rather quantifiable likelihood and value, where there aren't high-order effects and downstream consequences to consider, where the decision you're facing is repeated many times, and where a single bad decision won't ruin you.

When I drive to work, there are two routes I usually choose from. One of them is a reliable thirty minutes on country roads. The other is on the highway and usually takes only fifteen minutes but sometimes gets backed up and takes around forty-five minutes. Can expected value suggest which is the better option? Yes, but only if it's used with some care.

If I leave for work early enough that there's no real chance of being late for class or missing a meeting, then all I really care about is minimizing my commute time. In this case, expected value says the country road is a better bet if highway backups happen more than half the time; otherwise, the highway is better. For example, if backups happen 25 percent of the time, then my expected commute time on the highway is $0.75 \times 15 + 0.25 \times 45 = 22.50$, which is indeed less than the thirty-minute country road commute.

This is a reasonable use of expected value because it satisfies all the criteria listed above: I do the same commute often, I can quantify everything I need for both options, and there are no complex long-term effects to worry about. On the other hand, if there's a meeting I must attend in the morning and I'm cutting it close, I ditch the formula. It's too hard to quantify the costs of being late, and being habitually tardy has long-term effects that wouldn't show up in a case-by-case expected value calculation. When I am rushed, I usually take the reliable thirty-minute drive even if it means being a few minutes late, rather than taking the chance of a traffic jam. Sometimes the meeting is so important that I feel compelled to roll the dice on the highway. But when I do, it's not because I crunched the numbers and decided it's worth the risk. It's because I just went with my gut. Part of being good at math is knowing when not to use it.

Another nice setting for expected value is taking multiple-choice tests. Until a few years ago, the SAT included a quarter-point penalty for getting a question wrong. Since each question had five choices, the expected value for someone who randomly guessed on each question was zero: $(1 / 5) \times 1 + (4 / 5) \times (-0.25) = 0$. That meant you did yourself no harm but also no favors by taking wild guesses whenever you had no clue. If you ruled out even a single answer, however, then the expected value became positive and you were wiser to guess. This wasn't obvious to everyone. Some people felt that eliminating one answer out of five didn't justify the gamble of getting a penalty for an incorrect guess. But one should ignore such feelings and trust the math. That's because standardized tests are as textbook a setting for expected value as it gets: the different outcomes are clear, all the probabilities and values are known, a single bad gamble won't bankrupt you, and there are no overlooked complexities lurking.

The SAT has since done away with penalties, so now you should guess even if you can't eliminate any answers. But there are plenty of test and test-like situations you might face, including at some job interviews, where a quick expected value calculation to guide your guessing strategy is helpful.

In the next chapter, we'll see how to protect ourselves from the risk of ruin that a naive application of the expected value formula overlooks.

5

How to Handle Risk

n 2015, *The New Yorker* ran an article titled "The Really Big One" with a foreboding subtitle: "An earthquake will destroy a sizable portion of the coastal Northwest. The question is when." The article cites a scientific estimate that there's a one-in-ten chance this devastating 9.0-magnitude earthquake will strike in the next fifty years. This was not a pleasant article for residents of the Pacific Northwest to read. And it raises many difficult questions. If you live there today, should you plan to move away? If so, when? If you stay, should you pay for an expensive retrofit of your home?

When COVID vaccines were first made available at the end of 2020, there was some concern among the public that the accelerated pace of development meant the vaccines were too risky to take right away. In the ensuing months, data surfaced suggesting a possible link between the vaccine and a rare heart condition. Was the vaccine worth the risk? And what precautions should you take when the next dangerous pandemic takes root?

Your old college roommate texts out of the blue with some news: they're launching a new tech startup, and you can invest now to get in on the ground floor. The pitch sounds intriguing. This could be the next TikTok, you tell yourself. But is it worth taking money out of your savings to gamble on this venture?

Risk comes in many flavors, and so does the mathematics of risk management. There are entire textbooks written on the subject, brimming with lengthy equations devoured by insurance executives and hedge fund managers. But these quantitative elites aren't the only ones who face risk. We all do. The rest of us shouldn't leave our fates up to chance. We can all use math to get a handle on risk. We just need the math to be made a lot simpler and easier to use. This chapter is my attempt to democratize the mathematical tools of risk management.

We'll start by seeing how two concepts familiar from chapter 3, variance and covariance, play an important role. Then we'll meet another math concept, the logarithm, that helps us reason more carefully about money. Think of these as appetizers in the mathematical meal we're cooking. Then we'll be ready for the two main dishes. The first is weighing options when risk is involved, and the second is deciding how aggressively to invest in opportunities that might not pan out.

Look Beyond Averages

In the 2021–2022 season, the five highest-paid players in the NBA—LeBron James, Steph Curry, Kevin Durant, James Harden, and Russell Westbrook—each earned over $40 million. But how much did most players make? The average NBA salary that season was $8.5 million. However, this one number, the average, doesn't capture the whole story. If we do what Francis Galton did with ox weights at the Plymouth fair in chapter 3, repeatedly tossing out the highest and lowest number until the middle value is reached, we find a median salary of $4.3 million. That means half of all NBA players earned less than $4.3 million, despite the average salary being nearly double that. That's because the big names at the top of the salary scale bring the average way up, whereas for the median those superstar contracts are counter-balanced by those of bench players at the bottom that you've probably never heard of.

Here's another way to think about the question of what "most" players earn. Group salaries into those under $1 million, those between $1 million and $2 million, those between $2 million and $3 million, and so on, all the way up to Steph Curry in the $45–$46 million range. Out of all forty-six of these salary ranges, the one with the most players, containing nearly 20 percent of them, is the second smallest: $1–$2 million. If you randomly selected a player in the NBA that season, this is the most likely salary range they would be in. That's still a great salary, but it's a lot less than what people typically envision when they think of the NBA.

Grouping salaries into evenly spaced ranges like this creates something called a histogram. These are very useful because they convey the full distribution of numbers and put them into a statistical framework. For example, the first bucket in our histogram has 13 out of the 427 total players, so an NBA player that season had a 3 percent chance of earning less than $1 million. The first ten buckets have just under three-quarters of the players, so there was a little over a one-in-four chance that any individual player got paid more than $10 million.

Moving past averages to distributions and histograms takes us a big step closer to understanding risk. If you're considering a major career change, don't just look at average salaries in the new profession; try to find the percentage of workers in different salary ranges. An average of $180,000 is of little relevance if 90 percent of people in the profession make less than $80,000. The novelist Mary Adkins recently ran a survey of nearly 1,500 published authors and found the average advance for a debut book was $57,000. But the median advance was far less, only $25,000. That means half of first-time authors who were fortunate enough to land a contract with a publisher earned an advance of less than $25,000. (And alas, many books don't earn royalties beyond the advance.) This point about averages and distributions isn't just a matter of money. COVID symptoms lasting on average around a week is cold comfort to the unlucky 6.5 percent or so whose infections resulted in long COVID symptoms that lasted several months or more.

Lifespan offers another fascinating example. For much of human history, average life expectancy at birth was around twenty-five years. Throughout the last millennium it barely inched upward, until the twentieth century when it shot up from thirty years to just over seventy. It's tempting to conclude from this that being elderly is a recent phenomenon. But that's simply not true. That's an illusion cast by averages.

Many people in the past lived long lives. Octogenarians make routine appearances in the historical record. Prior to the twentieth century, however, infant and childhood mortality rates were very high. This kept the average lifespan quite low. Those who survived to adulthood had lifespans much closer to today's than you might guess. In medieval England, boys born to landowning families lived to only thirty years on average. But the boys who made it to adulthood lived to fifty years on average. Health risks in the past were heavily concentrated in one's youthful years. While modern medicine has afforded us better outcomes at all ages, its biggest impact has been helping us survive to adulthood.

If we had histograms throughout the centuries showing what percentage of people lived to what ages, this nuanced story of longevity would leap out at us. We'd see people living to all ages throughout the centuries. And we'd see that prior to 1900 there were high percentages who died in their youth, but as the twentieth century progressed, these early deaths spread out across the other life stages. A sharp decline in the large percentage who died in their first twenty years led to modest increases in the percentages who died in their thirties, forties, fifties, and so on.

Histograms are wonderful tools, but in practice you seldom have access to the fine-grained data needed for such a detailed numerical breakdown. That's OK. You'll still think more clearly about risk by having histograms in mind, even if you don't have the numbers at hand. Histograms also help us better understand a certain statistical notion that plays a surprisingly large role in risk: variance.

When discussing predictions in chapter 3, I described variance as a measure of how spread out a collection of numbers is. That's not wrong, but it's also quite vague. Here's a more precise description. First, compute the average value of the numbers in the collection; call this the *center* of the collection. The variance is then the average distance from the numbers in the collection to this center. (Technically, you need to square all these distances, but let's not worry about that mathematical subtlety.) Histograms help us unpack this mathy definition.

Histograms are usually depicted visually as a bunch of adjacent vertical bars, almost like a cityscape in silhouette, where the heights convey how many data points lie in each range. I'm sure you've seen these in the news. Variance is the weighted average distance of the bars from the center, where each is weighted by its height. In terms of cityscapes, variance captures the extent to which there are tall buildings far from the city center. In terms of probabilities, it captures the extent to which there are probable values far from the expected value. Yes, that's the same expected value we studied in chapter 4. This is what connects variance to risk. Expected value, as you recall, says what to expect on average. Variance says how likely something is to defy your expectation. And things not turning out the way you expect is a major source of risk. Just ask Sam Bankman-Fried.

To see how variance signals risk, let's walk through some examples.

Imagine you're in the process of refinancing your mortgage. Your broker says there's a fancy new AI system for predicting mortgage rates. "It's indicating that rates will drop two percentage points in the next few months, so we should wait to refinance." You ask how it knows rates will drop, and the broker explains that it considers a bunch of scenarios and for each one works out what the rate would be. "The two-percentage-point drop," your broker continues, "is the average across all these scenarios."

The smart response would be to ask how large the variance is. If it's small, then most of the scenarios result in mortgage rates close to the expected two-point decrease. You can then safely delay refinancing with confidence that you'll very likely get a better rate by doing so. On the other hand, if the variance is big, then there's a greater chance the rate ends up far from the expected value. With large enough variance, the rate could even go up instead of down—which would make waiting to refinance a mistake. The higher the variance, the higher the risk in your plan to postpone.

Situations like this are not merely hypothetical. On March 20, 2023, the Intergovernmental Panel on Climate Change (IPCC) released the sixth and final installment of its almost decade-long analysis of climate change. This installment alone runs to nearly eight thousand pages and draws from the work of hundreds of scientists. Among other things, it analyzes a number of scenarios and finds that a warming of 1.5 degrees Celsius above preindustrial levels has more than a 50 percent chance of occurring by 2040. An increase of that magnitude, the report estimates, would leave nearly a billion people in drought conditions and result in a 14 percent decrease in biodiversity, a 24 percent increase in the number of people exposed to flooding, and a further 80 percent decline in coral reefs, to name a few of the dire consequences.

Those are terrifying statistics. And even if one accepts that level of warming as our fate, additional risks abound. There is a risk that 1.5 degrees Celsius strikes earlier than 2040. And there is a risk that its impacts are greater than anticipated. These risks have two overarching sources, both of which boil down to variance. First, the IPCC predictions are just that: predictions. The IPCC qualifies them with confidence scores indicating how likely it believes they reflect what will transpire. The less confidence the IPCC has in an estimate, the wider the range of possible outcomes—that is, the greater the variance. Second, the climate of tomorrow is shaped by our actions today. The IPCC's 50 percent chance of 1.5-degree warming by 2040 is based on a range of scenarios capturing the different courses of action society could take. The more variance in our actions, the less precisely we can pin down what will happen and when.

This is even more clear when we look further into the future. The IPCC estimates that in a high-emission, carbon-intensive scenario, temperatures will have risen by the year 2100 to somewhere between 3.3 degrees and 5.7 degrees Celsius above a preindustrial baseline. (For context, the last time we were more than 2.5 degrees Celsius above that baseline was over three million years ago.) Notice that for a prediction that far out, the IPCC does not even attempt a single number and instead offers a range, reflecting the uncertainty involved. This isn't a guarantee that global warming will be in this 3.3-to-5.7-degree range. Rather, the range helps convey that there is a distribution and gives a sense of its variance.

Remember the terrifying earthquake warning facing the Pacific Northwest that opened this chapter? In the past ten thousand years, there have been forty-one subduction-zone earthquakes in the region. These are the devastating quakes resulting from one tectonic plate sliding underneath another. The *New Yorker* article explains that "if you divide ten thousand by forty-one, you get two hundred and forty-three, which is [the] recurrence interval: the average amount of time that elapses between earthquakes." Do these quakes occur like clockwork precisely every 243 years, or is there a large variance in the timing between them, causing them to bunch up and spread out at unpredictable intervals?

The last such earthquake was around 9:00 p.m. on January 26, 1700. That's 325 years ago from the time this book was published, which is already eighty-two years longer than the average interval. We don't know the exact date of the quake before that, but it's believed to be around the tenth century. That means roughly 750 years elapsed between that quake and the subsequent one—triple the average interval. Evidently there is considerable variance in the time between quakes. Maybe we'll be lucky and the next one won't be for another 750 years. Maybe we'll be unlucky and it's right around the corner. If we knew when it would strike, it'd be a heck of a lot easier to prepare for it. Alas, the variance here means the specter of a devasting quake lingers over the Pacific Northwest with terrifying unpredictability.

Risk is a tricky concept to pin down. If pressed to define it in mathematical terms, I'd say it's the likelihood of something bad happening. This could be a specific bad thing: earthquakes pose a risk of striking; investments have a risk of tanking; changing careers has a risk of not earning as much as you expected; smoking increases your risk of cancer; driving drunk increases your risk of getting into a car accident; global warming increases the risk of extreme weather events. Or it could be something bad in general: it's risky to eat unhealthy foods, drink too much, go without insurance, or insult random people on the street. We'll spend the rest of the chapter exploring how to manage the risks we face in life. The main lesson from this section is that we must first change our mindset from focusing on averages and expected values to thinking of entire distributions. A common source of risk is the occurrence of the unexpected, for that can lay waste to our best-laid plans and predictions. Variance, which measures how spread out a distribution is, conveniently gauges the extent to which we should expect the unexpected.

Understand Diversification

You've no doubt heard that diversification is a good way to reduce risk in financial investments. It's worth digging into the math of diversification to understand how it works, how to do it effectively, and when not to do it. This will help not just in matters of money but in a range of situations you might face that involve risk.

First, how it works. Diversifying investments is a form of averaging. If you put half your money into Starbucks and half into Tesla, then your portfolio's performance will be the average of the performance of these two stocks. This reduces risk by lowering the odds that your investment tanks. To see how, let's pretend for the sake of simplicity that on the time scale you're considering for your investment, each stock has a 50 percent chance of dropping and a 50 percent chance of rising. If both stocks go up, their average goes up. If both go down, their average goes down. And if one goes up while the other goes down, by comparable amounts, their average stays even. In one of these cases you make money (up-up), in two of them you don't make any but you also don't lose any (up-down or down-up), and in only one of the four cases do you lose (down-down). Your diversified investment therefore has a 25 percent chance of dropping, much safer than the 50 percent chance you'd face if you bought only one of these two stocks.

However, there's a catch. In the above calculation, I sneakily assumed that the two stocks move independently. That's a bit like saying that if you have two kids, then there's only a 25 percent chance they're both boys, because the possibilities are girl-girl, girl-boy, boy-girl, and boy-boy. Yes, that's usually the case—but not when they're identical twins. For identical twins, the possibilities are girl-girl and boy-boy. The odds of having two boys jumps up from 25 percent to 50 percent. Stock movements usually fall somewhere between these two extreme situations: they're not totally independent, but they're also not totally synced up like the sex of identical twins.

It would be nice if we had a mathematical concept that measures how linked two stock prices are. Fortunately, we do, and it's one we saw already in chapter 3: covariance. Remember that this is the tendency of two sequences of numbers to move in the same direction at the same time. It tells us when two marble-guessers tend to overestimate and underestimate the same jars, and it tells us when two stocks tend to go up (or down) on the same days. Covariance also indicates the extent to which diversification reduces risk: the greater the covariance among the investments in a portfolio, the more illusory the diversification becomes and the less risk is lowered.

Suppose you want to invest in a green energy company, because it's something you believe in and want to support but also because you anticipate growth in that sector of the economy. You find a promising company doing residential solar panel installations that seems like a good choice, but there's always a chance something goes wrong: business mismanagement, inability to gain market share, a corporate scandal, or some other unforeseen impediment to success. These problems are usually spread out within an industry, so investing in two or more residential solar energy companies lowers your risk of financial failure. You can diversify even further by investing in some companies that do residential solar and others that do non-residential solar, because there is surely some degree of covariance among all residential solar energy companies. But there's a chance the entire solar industry struggles, so I'd recommend diversifying more broadly across the green energy industry by investing in some solar companies, some wind, and some geothermal.

At that point you might say, "But wait, all green energy companies have covariance to an extent, so why not go even further and add some oil and natural gas companies to the mix?" Some investors think that way and act accordingly, but this loses sight of your original plan to support the green energy economy while tapping into the growth you expect it to exhibit. The goal isn't to diversify away all conceivable covariance; it's to build a portfolio that focuses on your areas of investment interest and that smartly diversifies within those areas.

One convenient way to tap into the advantage of an average when investing is with a mutual fund or an ETF that tracks an index or sector of interest. For example, there are green energy ETFs whose performance reflects an average across a range of companies in the industry. Funds like these aren't chosen to minimize covariance, but they typically have a wide enough range of stocks that they are well-protected from risk. You can diversify further by getting an ETF that tracks one of the broad market indices we saw in chapter 2. The first-ever ETF, introduced in 1993, tracks the S&P 500. It is currently the largest ETF and a popular choice among many investors. An investment strategy that simply tracks the market average might seem too obvious and easy to be worthwhile, but it works remarkably well. If you put $10,000 into the original S&P 500 ETF back in 1993, it would be worth $185,000 in 2023. That amounts to an impressive annual return of around 10 percent across this three-decade span.

For most of us, this kind of broad-swath market average approach to investing is an excellent strategy. It is so heavily diversified that your risk is very low, and historically its returns have been very strong. There have certainly been painful periods when the market as a whole stumbled, but those who held on to their investments and rode out these dips were rewarded for doing so in the long run.

That said, some ambitious investors don't want to match the market; they want to beat it. One such investor is a household name: Warren Buffett. As of August 2024, his net worth of $142 billion makes the ninety-four-year-old Buffett the sixth-richest person on the planet. Here's what he once said about tracking the market versus aiming to beat it: "By periodically investing in an index fund, the know-nothing investors can actually outperform most investment professionals. If you are a know-something investor, able to understand business economics, and can locate five to ten sensibly priced companies that possess important long-term competitive advantages, conventional diversification makes no sense for you." He made this point even more bluntly on another occasion: "Diversification serves as a protection against ignorance. . . . It's a perfectly sound approach for somebody that doesn't know how to analyze businesses."

Buffett is not a reckless, high-risk investor. When he disparages diversification here, he's not saying investors should throw caution to the wind. He just has a different approach to managing risk. Buffett likes to think more about individual businesses than entire industries or sectors of the economy. What are the risks an individual business faces that could hinder its performance? Incompetent management, a new competitor in the market, a trendy product that may fall out of favor with customers, things of that ilk. Buffett doesn't view these as mere matters of chance to be avoided through averaging across businesses. He views them as foreseeable problems one can avoid by thoroughly analyzing the business fundamentals of the companies one is considering investing in—and by being willing to wait long enough for the right opportunity to arise. Averaging helps reduce bad luck, but Buffett tries to take luck out of the equation by doing his homework.

Of course, even with this focused approach, Buffett doesn't put all his eggs in one basket. Notice that he suggested choosing five or ten stocks, not just one. Buffett and investors like him reduce the risk of each stock they purchase by looking very closely at the business it represents. But they also reduce risk across their portfolio by choosing stocks that have low covariances. Buffett famously has been a big investor in Coca-Cola and in GEICO. These are very different companies in very different industries. It's hard to imagine their performance being tied in any real manner, which means they are a stock pair with precious little covariance.

Most of us amateurs are not qualified to attempt the kind of small portfolio investing that Buffett recommends. Speaking for myself, I'm much closer to the "know-nothing investor" than the "know-something investor" he referred to. Nonetheless, a valuable lesson here is that there are two main ways to reduce risk in investing: (1) build a large portfolio, perhaps with the aid of ETFs or mutual funds, to average out any unexpected stinkers; (2) do extensive research into the businesses you invest in to root out any foreseeable problems.

These are both effective ways to reduce risk, but it's practically impossible to embrace them fully at the same time. The more stocks you toss into your portfolio, the less carefully you will be able to scrutinize each one. And the less likely each one is an amazing, unique opportunity you've been waiting years for, rather than just a good-enough option at the moment. Think of these two risk-reduction approaches as ends of a spectrum. Buffett leans heavily toward the second, while most amateurs would do better to lean more toward the first. But with either approach, you get more bang for your diversifying buck by thinking hard about covariance and building a portfolio that doesn't have too much of it.

In the end, keep in mind that diversification is a form of averaging. And as we learned with racist juries in chapter 3, covariance kills the efficacy of an average.

Let's now look beyond investing. Experts have recently been warning that AI will disrupt many industries by rendering some human workers obsolete. To reduce risk in your career choice, you should think carefully about which jobs are most threatened by AI automation. One strategy is to look up employment forecasts published by labor economists. When doing so, you can take a page from chapter 3 and use a weighted average to combine the forecasts you find. Then aim for the jobs anticipated to be the safest from AI upheaval.

Another strategy is to diversify. If two jobs involve similar tasks, then when AI comes for one of them, it will likely come for the other as well. Try to build skills that use very different parts of your brain and prepare for careers that rely on this wide range. That way, even if AI supplants some of what you do, it might still lag far behind in other areas. Covariance is technically a numerical measure, and no numbers are involved here. But I nonetheless find it helpful to keep the concept of covariance in mind in situations like this. To insulate yourself from the risks posed by AI and automation, you want to diversify your skill set. That doesn't just mean learning more skills. It means learning skills that are as different from each other as possible—skills with minimal "covariance" between them, so to speak.

To take our analysis of risk to the next level, we need to better understand the impact money has on our well-being, both physical and mental. This requires a brief detour into the realm of psychology. We won't be leaving math behind while doing so. Quite the opposite; we'll be confronting a concept you probably saw in high school math class that probably didn't make much sense at the time: the logarithm. We'll see it in a whole new light here.

Think Logarithmically

Barbara Mellers is a University of Pennsylvania psychology professor whose work focuses on decision-making. For many years she has worked closely with her UPenn colleague Philip Tetlock, author of the bestselling book *Superforecasting*. They founded one of the main forecasting tournaments mentioned in chapter 3 that Molly Hickman's group, Samotsvety, dominates. Mellers and Tetlock helped establish these tournaments to train forecasters and to study the traits the best forecasters have in common. One of their remarkable findings, and an impetus for the formation of Samotsvety, was that crowdsourced predictions, where amateurs band together, often outperform expert predictions. All the more reason to combine lots of different predictions with a weighted average—the mathematical version of wisdom of the crowd.

Mellers recently pivoted from predictions to tackle an age-old question: Can money buy happiness? Specifically, she wanted to know how much a person's happiness increases as their salary increases. This is a question that had been researched before, but the findings are confusing. Most data suggests that, as you'd expect, happiness steadily goes up with salary. However, an influential study from 2010, led by the late psychologist and Nobel Prize–winning economist Daniel Kahneman, found that happiness sharply tapers off around a salary of $75,000. In 2023, Mellers teamed up with Kahneman and a happiness researcher at Wharton, Matthew Killingsworth, to unravel this mystery.

Mellers and her collaborators revisited the 2010 data and found an important overlooked issue: what was being measured there wasn't so much happiness as it was unhappiness. In this light, the plateau makes more sense. The burdens of poverty diminish as one's salary rises to around $75,000. At that point, a rising salary does continue to increase one's happiness, but it no longer decreases one's unhappiness. While in retrospect it's rather obvious that people's happiness generally goes up as they earn more money, what's interesting is the *rate* at which it does so. To explain this, we first need to understand the logarithm.

The easiest way to think about the logarithm is that it counts the number of digits after the leading digit: $\log(10) = 1$; $\log(100) = 2$; $\log(1,000) = 3$; $\log(1,000,000) = 6$. Well, this works if the leading digit is one and the rest are zeros, but that's enough to give you a sense of how the logarithm works. For instance, since 50 is between 10 and 100, $\log(50)$ is between $\log(10)$ and $\log(100)$, meaning it is between 1 and 2. A crucial property of logarithms is that no matter what number you start with, the log goes up the same amount whenever you add a digit. The increase from $\log(5)$ to $\log(50)$ is the same as the increase from $\log(50)$ to $\log(500)$ and from $\log(5$ million$)$ to $\log(50$ million$)$.

Many things we experience in life are measured using a logarithmic scale. That's because we're often more interested in the number of digits a quantity increases by—a CEO's compensation package rising from six to seven or eight figures, for example—than the sheer amount it increases by. If the energy of an earthquake increases by three digits, then we say its magnitude went up by two. So a 7.0-magnitude quake has a thousand times more energy than a 5.0 quake and a million times more energy than a 3.0 quake. Other examples of logarithmic scales include the pH scale used to measure acidity, the magnitude system used to measure the brightness of stars, and the decibel scale used to measure loudness.

A logarithmic scale captures what we experience in these settings. The loudness of a refrigerator is 50 dB, a sewing machine is 60 dB, a coffee grinder is 70 dB, a doorbell is 80 dB, a blender is 90 dB, a snowblower is 100 dB. Most people would say this seems reasonable, that each item in this list is roughly the same amount louder than the item preceding it. The energy of the sound waves created by these items increases exponentially, but the logarithm underlying the decibel compresses this exponential growth back down to a scale that more closely matches our perception of loudness.

Economists and psychologists have long suspected that the impact money has on us is also logarithmic. Winning a $95 million jackpot like the one we heard about in Texas would certainly be thrilling, but do you really think winning a $190 million jackpot would be *twice* as thrilling? Or have twice the boost to your happiness and quality of life? Don't count on it. Winning $95 million would transform an ordinary person into a fabulously wealthy one; winning another $95 million would make them even wealthier, but it wouldn't transform their life in the same way as the first $95 million. Logarithmic impact here would mean the difference between a $1 million jackpot and a $10 million jackpot is about the same as the difference between a $10 million jackpot and a $100 million jackpot. I've never experienced any of these, but at first blush that seems about right.

Here's a more relatable experience. If your hourly wage goes from twenty to forty dollars, your life will get a lot easier. Going from forty to sixty dollars is still an improvement, but it doesn't feel like as big a one, even though both raises are of the same amount. The logarithmic theory of money says you'd have to go from forty dollars all the way up to eighty dollars to get the same impact. That's because what I told you about logarithms counting digits comes from a more general property: multiplying a number by a fixed amount has the same impact on the logarithm no matter what number you start with. Adding a digit is multiplication by ten, so adding a digit increases the logarithm by the same amount no matter what number you add the digit to. Doubling, as we did from twenty to forty, then forty to eighty, is multiplication by two, so doubling increases the logarithm by the same amount no matter what number you double.

In his eighties, Warren Buffett's net worth rose from $50 billion to $100 billion. The logarithmic theory of money suggests that this meant about as much to Buffett as when his net worth rose from $50 million to $100 million in his forties, and from $5 million to $10 million in his thirties. A million dollars does not mean to Buffett the billionaire what it did to Buffett the millionaire. When it comes to money, we should focus on digits and doubling, not on dollars.

In settings ranging from household income to per capita GDP, empirical evidence (including the 2023 Killingsworth-Kahneman-Mellers study) has mostly supported the hypothesis that the socioeconomic and psychological impacts of money are roughly logarithmic. It's far from a perfect characterization, but it's close enough to capture the diminishing returns we experience from increased wealth. Think of the logarithm as a helpful heuristic for translating between money and happiness, even if it's not a precise formula for doing so.

This logarithmic perspective of money plays an important role in the formulas we'll explore in the next two sections. Even before we get there, it helps us avoid some common missteps. I believe we tend to overestimate how much better our lives would be with more money, because we tend to think linearly. Twice as much money won't make you twice as happy. Exponential growth is required for your happiness to rise at a steady rate. People often colloquially use the term "exponential growth" for anything that increases quickly, but here I mean it in the strict technical sense. Exponential growth isn't just rapid growth; it is a type of growth that perpetually accelerates. So, for your happiness to keep increasing without slowing down, it's not enough for your wealth to increase without slowing down. Your wealth would need to keep growing at faster and faster rates.

If you're in a happy situation, for instance a job you enjoy, it's often better to stick with it than to risk losing that happiness for the chance of more money. The extra money likely won't impact you as much as you expect. That's certainly not a black-and-white rule; each case should be analyzed on its own terms. But it's worth keeping in mind.

Romer's Rule-of-Thumb Recipe for Risk

Paul Romer won the Nobel Prize in Economics in 2018 for work that we'll discuss in chapter 8. When I told him about this book, he was happy to hear there'd be a chapter on risk management—especially because of what transpired during the pandemic. In the early days of the outbreak, Romer had advocated for an aggressive COVID testing regime funded by the federal government. He argued that lockdowns risked economic collapse, while returning to a pre-pandemic way of life without a robust testing protocol in place risked the deaths of millions more Americans. Supplying enough tests to safely reopen the country seemed like a win-win plan, saving both lives and dollars in the long run. Yet it was largely ignored by the powers that be. Government officials did not price risk appropriately, in Romer's view.

As the pandemic raged on, Romer shifted his efforts. Frustrated with the resistance many expressed toward the new COVID vaccines, he tried to help the public understand that these vaccines were overwhelmingly worth the risk. Despite their side effects and the disappointing realization that they offered only partial protection and didn't completely prevent contagion, they were hugely beneficial both for the individual and for the population at large. Yet the resistance remained. Once again, Romer was exasperated with the way risk was mismanaged—this time at all levels of society. As Liv Boeree, a British poker champion we'll meet at the end of this chapter, has written, "If the mess of public confusion and poor leadership surrounding the coronavirus pandemic has taught us anything, it is how poorly equipped we are to navigate risk and uncertainty."

Romer shared with me some thoughts on the mathematics of risk that I'd like to share with you here in simplified form. To start, he notes that it's not just our actions that carry risk. Our inactions do, too. We should take both forms of risk into account. Unfortunately, we seem to have a cognitive bias that weighs more heavily the risks associated with our actions than those associated with our inactions. This is a particularly vexing issue with vaccines. As we saw during the pandemic, many people chose to avoid the risks of a novel vaccine, even if doing so meant an increased risk of infection, hospitalization, and fatality from a novel virus that had already killed millions and disabled even more. Romer ran the numbers repeatedly, and even with the most conservative of estimates, the risks from the virus were orders of magnitude greater than the risks from the vaccine. Yet many acted as though the opposite were true.

The gist of Romer's recommendation for making a decision in the face of uncertainty is that we should do a standard cost-benefit comparison of our options, being careful to include both actions and inactions. But there's one crucial addition: the costs of the main risks should be calculated and included in the cost-benefit comparison. Let's explore what this means and see how Romer suggests we do it.

Some costs are clear, such as paying the out-of-pocket charge for a medical treatment. Other costs involve some degree of uncertainty, such as whether you'll be ticketed for speeding and how much the ticket will be if you are. These uncertain costs are the risks. The way to price a risk is to decide how much you'd be willing to pay to avoid it. This can be done for risks of all types, not just financial ones. For instance, you can price the risk that a medical procedure goes wrong and leaves you with permanent damage. Insurance companies do this sort of calculation all the time. The challenge is that you don't have an army of actuaries to crunch the numbers the way the insurance companies do.

Thankfully, Romer offers a convenient rule-of-thumb recipe for this thorny task of pricing risk. First, figure out a dollar amount reflecting the harm the risk could cause. We'll soon see an example where the harm is missing work and the dollar amount is the lost income. Sometimes you'll need to translate a nonfinancial harm into a dollar amount, which is tricky. It helps to switch perspectives from deciding how much the harm would cost to deciding how much you'd be willing to pay to avoid that harm. Romer suggests sticking to a crude power-of-ten ballpark: Would you pay $10,000 to avoid the harm? $100,000? $1,000,000? $10,000,000? Next, multiply this dollar amount for the harm by the probability that the harm occurs. This yields the *expected loss* for the risk. The third and final step is to double this expected loss.* Doing so gives Romer's rule-of-thumb cost for a risk, allowing you to bring risks into any cost-benefit analysis.

Imagine there's a virus spreading rapidly in your area. It doesn't pose any life-threatening risks or cause any long-term harm, and nine out of ten people who get it are fine after twenty-four hours. However, the remaining 10 percent end up sick in bed for a week. You start to feel funny, and you're out of sick days at work, so you rush to the doctor. A test shows you have this virus.

---

*Here's the idea behind this doubling of the expected loss. The logarithm of the financial outcome when the risk occurs, times the probability of the risk, plus the logarithm of the financial outcome when the risk doesn't occur, times the probability the risk doesn't occur, is called the *expected utility*. This is how much you expect to end up with, on average, measured logarithmically to capture impact. Exponentiating converts this back to a dollar figure and provides the amount of money you should be happy to end up with in lieu of taking the gamble on the uncertain risk. The difference between this amount and the full amount when the risk doesn't occur is the price you should be willing to pay to avoid the risk. Some mathematical mucking around reveals that this difference is approximately twice the expected loss, at least when the probability of the risk is relatively low—which it typically is for the risks that enter cost-benefit calculations.

The doctor says there's an easy treatment that will prevent you from getting the weeklong illness, but it's not covered by your insurance and costs $200. Should you do it?

Your first thought is no. There's a 90 percent chance you'll be fine tomorrow without the treatment, so why spend so much money on something you probably won't need? But then you remember Romer's lessons on risk—perhaps you're succumbing to the cognitive bias of leaning too heavily toward inaction. You decide to do the math. Let's say you earn $5,000 each month. The cost of missing work for a week is a quarter of this, so $1,250. There's a 10 percent chance you get the weeklong ailment, so the expected loss is $0.1 \times \$1,250 = \$125$. Doubling this yields a rule-of-thumb cost of $250 for this risk. Your intuition was wrong: you should pay for the treatment, even though you'd most likely be fine tomorrow without it. And this $250 figure is an underestimate since we accounted for only the lost income from being sick, not the week of physical discomfort.

Now let's try converting a nonfinancial harm into a dollar figure. During the pandemic, Romer wondered what people would pay to prevent long COVID. To find out, he gave the students in one of his economics classes a hypothetical scenario: "Imagine I'm doing a medical research project on long COVID and we found a way to inflict this condition on people; this helps us learn about the disease and maybe one day discover an effective treatment, but we need volunteers. How much would I need to pay for you to get long COVID and be part of my medical study?"

Perhaps unsurprisingly, their initial answer was that there is no amount of money that would compensate for accepting this potentially incurable condition. "It's crazy to put a dollar amount on something like this," one student objected. "It's not just crazy, it's unethical," another added. But soon the nut started to crack. "I'd do it for a billion dollars," one student offered. "Yeah, well, of course for a billion dollars I'd do it, too," others admitted. Once they had this concrete number in front of them, it was easy to lower the bar. Ultimately, the room settled on $10 million as a reasonable ballpark cost of long COVID. This was early in the pandemic; it's possible that we'd get a lower number now that there have been more stories of people recovering from the ailment. But let's put ourselves back in those terrifying days when suffering and fear were rampant, and stick with $10 million.

Assigning a cost to long COVID allows us to proceed with Romer's recipe. Around three-quarters of Americans have had at least one COVID infection, and around 6.5 percent of these individuals ended up suffering from long COVID. That means if you lived in the US during the pandemic, you faced roughly a 5 percent chance of getting long COVID. We thus find an expected loss of 0.05 × $10,000,000 = $500,000, which when doubled yields a $1 million rule-of-thumb cost for this risk. Imagine you were working from home during the pandemic and so carefully locked down that your chance of getting COVID was essentially zero, but then your boss begged you to return to in-person work and asked how much of a cash bonus you'd need to tolerate the risk. If, as a crude estimate, you ballpark that returning to the office would increase your chance of getting COVID from zero up to the general American population average, then Romer's math says you should ask for a $1 million bonus to offset the risk of long COVID. That may sound like a lot, but a 5 percent chance of months, years, or even a lifetime of a mysterious, potentially untreatable, and often debilitating ailment is a lot, too.

Sometimes it feels uncomfortable putting a price on your health like this and using numbers so subjectively—perhaps even reminiscent of the EPA's disturbing valuation of human lives mentioned in chapter 1. But I believe there's a lot less risk in using numbers to help make decisions in your own life than there is in a large organization using numbers to make decisions that impact many other lives. That said, it's perfectly OK to draw a line in the sand and declare that some things in your life simply defy quantification. I love math, but I don't use it for everything I do. Just make sure any decision to forgo numbers is because you genuinely don't think numbers are appropriate in that setting— not because finding good numbers is too hard.

This brings up another cognitive bias worth being aware of. We tend to avoid doing calculations for fear of getting the numbers wrong. But the decisions we make without any calculations are often far worse than the ones we make even with very rough estimates. A motto Romer and I recommend to help people overcome this bias is "close enough for jazz," which basically means "good enough." Whisper this to yourself when you feel intimidated trying to come up with ballpark estimates; it's a simple reminder to jump in and try rather than letting the specter of unattainable perfection hold you back. Just be careful not to get carried away quixotically quantifying everything under the sun. Like most things in life, numbers should be used in moderation. Importantly, numbers should reinforce our thinking—not replace it.

I like to view numbers, and the calculations they enter into, as signposts that help direct us and keep our intuitions from going too far astray. When it comes to managing risks and making decisions in the face of uncertainty, intuition is especially prone to error. Romer's cost-benefit recipe, and the formula we're about to encounter next, redirect us back onto the path of reason and rationality.

A Formula for Balancing Risk and Reward

Imagine you're one of three lucky fans at a WNBA game selected for an exciting opportunity. Each of you is given a $1,000 cash prize to start out with. Caitlin Clark, the unparalleled sharpshooter who broke just about every scoring record in college basketball, will take a bunch of shots from the half-court line. For each shot, you choose an amount to wager from your prize and guess whether the shot goes in or not. When you get it right, your prize money goes up by the amount you wagered; when you get it wrong, the wager is subtracted. The announcer begins by informing all three of you that Clark is a 60 percent shooter from that distance.

The first fan, having just read chapter 4, recognizes that by always betting on Clark sinking the shot, the expected value formula says more money will be made than lost. The fan doesn't want to miss any opportunity to cash in, so they wager as much as possible every time. The first shot swishes, and with that sweet sound the prize money rises to $2,000. Next shot, same strategy: all in on Clark sinking it. And she does. As Clark squares up for the third shot, the fan wagers all $4,000 in the prize pool on another made shot. The ball rims out, and the fan walks away empty-handed.

The second fan, having read chapter 4 more closely, recognizes the first fan's mistake: gambler's ruin. Even with favorable odds, if you bet it all every time, you're guaranteed to lose everything eventually. But what to do instead? Bet all but a dollar each time? After a few shots a miss is inevitable, then the pool would be down to one dollar. That wouldn't end the game, but building back up to $1,000 isn't likely, especially with more misses surely occurring along the way. Betting all but a hundred dollars each time wouldn't fare much better. The fan decides to try a different approach: just wager a hundred dollars on Clark making each shot. She ends up taking ten shots and makes six of them. The fan leaves with $1,200.

Now it's your turn. You're feeling confused and conflicted. Betting too aggressively as the first fan did is a terrible approach, but you suspect that the second fan was too conservative and squandered an opportunity to make more money. The announcer asks your bet, but your mind goes blank. How should you play this game?

While I fabricated this half-court contest, a very similar

one (mathematically speaking) took place back in 2016. Victor Haghani, a former hedge fund manager, invited a few dozen people to attend a talk. The guests arrived expecting a lesson on investing, and they got one—but not the kind they anticipated. There was no lecture taking place. There was just a game to play. As the attendees later found out, they were unwittingly the subjects of an intriguing experiment.

Each participant was told to visit a website where they were given twenty-five dollars in online funds to bet on the outcome of a simulated coin flip. They were informed that this coin was programmed to land heads 60 percent of the time. The rules for betting were identical to our half-court contest: wager as much of your prize pool as you like on each coin toss, double the wager if you win and lose the wager if you guess wrong. Participants were allowed to bet on as many coin tosses as they could fit into thirty minutes. The winnings for each person were capped at $250, but the participants weren't told this at the outset.

Haghani and a colleague wrote a paper describing this experiment. They were interested in it because there's a simple math formula that says exactly how to place optimal bets in a situation like this, a formula for balancing risk and reward. Haghani and his coauthor note that "notable investors such as Warren Buffett, Bill Gross and James Simons have all reportedly made use of" this formula. The experimental subjects gathered that day largely comprised "college age students in economics and finance and young professionals at finance firms," yet they fell far short of what could have been achieved. In the words of Haghani and his coauthor: "Our subjects did not do very well. While we expected to observe some suboptimal play, we were surprised by the pervasiveness of it."

The optimal strategy in this game is to consistently wager 20 percent of one's prize pool on a bet of heads (momentarily I'll explain where this number 20 percent comes from). Since the pool starts at twenty-five dollars, this means an opening wager of five dollars. If that first coin comes up heads, then the pool rises to thirty dollars and the next bet should be six dollars on heads. If the first flip instead lands tails, then the pool drops to twenty dollars and the second bet should be four dollars on heads. This system ensures that you place larger bets when you have more money to work with and smaller bets when you have less.

If all sixty-one participants had used a fixed percentage betting strategy like this and had chosen a percentage anywhere between 10 and 20 percent, then around fifty-eight of them—that is, all but three—would have hit the maximum payout of $250. In the experiment, only thirteen reached the maximum payout. Notably, seventeen participants—just shy of a third of the whole group—lost the entire prize pool and left empty-handed, like the first fan in our fictional half-court contest. For a game with favorable odds, this was quite the lesson on gambler's ruin.

Let's now take a look at the winning formula and the math behind it. In a betting game like Haghani's coin toss experiment and the half-court contest, write p for the probability of the more likely of the two outcomes. (In our two examples these outcomes are a heads and a made shot, and both have $p = 0.6$.)

Because there are only two possible outcomes each round, this probability must be at least 50 percent. The optimal strategy is to keep betting a fixed fraction of your prize pool, and this fixed fraction should be $2p - 1$. When $p = 0.6$, as in our two examples, this comes out to 0.2, or 20 percent.

For those interested in where this formula comes from, you can find a full mathematical treatment in a number of textbooks or in the Wikipedia article for "Kelly criterion." (This system of betting is called the Kelly strategy, named after the mathematician who discovered it in 1956.) You'll also find generalizations of the formula that go beyond the case we've discussed of even-money bets in which you either double or lose your wager.

The formula provides the optimal betting strategy in the sense that it maximizes the expected value of the logarithm of the winnings. Why the logarithm? Because what really matters is the impact the earnings will have on your life, not the raw dollar value of the earnings. And as we saw earlier, the logarithm conveniently measures money's impact. This is the same principle underlying Romer's recipe for dealing with risk. You don't need to compute any logarithms when using Kelly's strategy or Romer's recipe. But in both formulas, the logarithm is a friendly ghost in the machine making sure that the diminishing value of money we experience is properly taken into account.

There are two main challenges to applying the mathematically pristine Kelly strategy and its variants to the messy realm of investing. First, the possible returns on an investment range continuously; there isn't a simple win or loss like in the games discussed above. One workaround for this is to focus on the probability p of an investment going up, since it either goes up or it doesn't, but this overlooks the amount it goes up or down. The second problem is that in the stock market, or any other investment arena, we don't know p; we can only try to guess it. This causes most investors to add a margin of safety deducted from the mathematical optimum the Kelly strategy aims for. A common approach is to use the Kelly formula but divide the result by two, meaning invest half as much as the original formula suggests. This is called a half-Kelly. This halving of Kelly to incorporate some risk aversion is similar to the doubling Romer used in the final step of his rule of thumb for pricing risk.

There seems to be some debate over whether Warren Buffett really uses the Kelly formula, as Haghani and others have claimed. Either way, the spirit of the formula comes through strongly in his actions and words. One of Buffett's closest companions in the world of investing was the late Charlie Munger. A fellow Omaha native, Munger befriended Buffett in 1959 and served as the vice chair of Berkshire Hathaway from 1978 until his passing in 2023. But this official title understates the influence he had. As Robert Hagstrom wrote in his bestselling book on Buffett, "In every way, [Munger] functions as Buffett's acknowledged co-managing partner and alter ego." Munger shared Buffett's understanding that diversification brings investors up to the average while also dragging them down to it. To beat the market, one must limit the number of stocks one owns. Even as few as three stocks was enough for Munger, if he felt strongly enough about them. And this is where Kelly comes in.

The fewer stocks one owns, the bigger a fraction of one's

portfolio is invested in each. In other words, the larger a bet one is placing on each stock. The Kelly strategy thus says to lower your diversification and place large investments in a smaller number of stocks when you have high confidence in them. Munger gleaned this insight from his days (or, more likely, nights) at the card table: "I knew from being a poker player that you have to bet heavily when you've got huge odds in your favor." This is the essence of Kelly: as p rises, so too should the fraction of your wealth that you wager.

The idea of a fixed-percentage wager, the main component of the Kelly strategy as we have seen, was never far from Buffett's mind. "With each investment you make," Buffett once explained, "you should have the courage and conviction to place at least 10 percent of your net worth in that stock." The famously patient Buffett doesn't just invest in the best opportunities available at the moment; he waits until the right opportunity arises. If, as a ballpark estimate, Buffett waits until he finds a stock he thinks has a 60 percent chance of doubling at the time horizon he's considering, then the Kelly formula $2p - 1$ would say to wager 20 percent on it, just as we've seen with both Caitlin Clark and the Haghani coin toss experiment. Using a half-Kelly to reduce risk brings this number down to 10 percent, matching Buffett's suggestion.

We don't know if Buffett does explicit Kelly calculations like this. But we do know that his approach to investing, particularly his willingness to eschew traditional diversification to beat the market, reflects the spirit of the Kelly formula. Whether you follow Buffett's lead or take a safer, simpler approach by diversifying more heavily, remember that the more confident you are in an investment, the bigger a fraction of your portfolio you should devote to it. And that the less covariance it has with your other investments, the more the overall risk in your portfolio is reduced.

Putting Risk into Perspective

Before he got into baseball analysis and election forecasting, Nate Silver, the prediction pro we encountered in chapter 3, spent a few years earning a living as a poker player. He continues to play semiprofessionally today. In 2023, he finished eighty-seventh at the World Series of Poker Main Event. As impressive as that is, it pales in comparison with another poker pro who likes to think of life in probabilistic terms. Liv Boeree graduated from the University of Manchester with an honors degree in astrophysics, then decided to enter a few TV game shows to earn some cash while she figured out what to do next. One of these shows involved expert poker players coaching amateurs. She was one of the amateurs. This was Boeree's first exposure to the card game, but it immediately appealed to her mathematical mind. Five years later, she won the largest European poker tournament, only the third woman to ever do so, earning a prize of over a million pounds.

In an interview, Boeree explained that chess is a game of complete knowledge: both players see all the pieces on the board, and there's no hidden information. But in poker, you don't know what cards the other players hold or what cards will be drawn from the deck, so playing strategically requires working with probabilities. What are the odds that the opponent is bluffing? Or that the next card turned over is a heart? Boeree said that poker "becomes largely a science of being able to quantify these uncertainties." Over the course of her poker career, she trained herself to think in probabilistic terms, and she found that the insights gained from this mode of thinking reach far beyond the card table. One of these insights, which she discussed in a TED Talk, offers a helpful perspective of risk.

Boeree said that in poker she has seen unlucky events occur so often that they seem to defy their improbability. To understand why this kept happening, she studied her past games and realized that while each unlucky event on its own is unlikely, there are so many different ways bad luck can strike that the likelihood of *something* unlucky happening is quite significant. And the same is true in life more broadly. Rather than seeing this as a depressing realization that we're all doomed to suffer bad luck, she found this liberating: it means we shouldn't feel so guilty and responsible when bad things happen. We should take precautions to reduce the risk of negative outcomes in our endeavors. But when something inevitably does go wrong, it doesn't mean we were too reckless. Don't read too much into bad luck; just accept it as an unavoidable aspect of life. We'll hear more from Boeree on thinking probabilistically in the next chapter.

For now, go forth into a world rife with risk, emboldened by the concepts and formulas from this chapter. It was a lot of material, so let's quickly recap.

Averages are not the whole story; to better prepare for the

unknown, you should think about distributions and histograms. These tools help us understand variance, which measures how spread out your numbers are and in doing so tells you how much you should expect the unexpected. This is relevant for risk, which often takes the form of things not going according to plan. One way to reduce risk, particularly in investing but more broadly as well, is to diversify—and to do so judiciously by choosing options with minimal covariance. When making decisions around money, remember the logarithm: it's digits and doubling, not dollars, that matter most. When weighing different options in the presence of risk, try Romer's recipe for doing a cost-benefit analysis. Even if you don't remember or use the Kelly formula, remember the main idea behind it: the more likely something is to turn out well, the more chips you should place on the bet. And finally, don't blame yourself for bad luck. Infrequent things frequently happen. That's not a contradiction or a paradox; it's math.

6

A Tool for Thinking
More Clearly

Throughout her youth, Liv Boeree was plagued with anxiety. At age thirteen, she read of a girl who lost her hair to alopecia. For the next six months, Boeree counted the hairs in her brush, worried she would develop this relatively rare disorder. A dozen years later, she was convinced that she acquired an exceedingly rare brain-eating amoeba from a wakeboarding trip in Las Vegas. "The only thing eating my brain was my irrational anxiety," she later wrote in an essay published the year after she and her boyfriend won a tag-team event at the World Series of Poker. As Boeree turned to poker, she developed a fluency for probabilities that helped her tame her hypochondria. She credits an unlikely adviser: "The most powerful antidote to my irrationality came from a surprising source: an 18th-century English priest named Reverend Thomas Bayes."

Boeree is alluding here to a remarkable math formula discovered by, and named after, the Reverend Bayes. This formula helps us grapple with uncertainty by handling likelihoods and evidence in a way that our brains unfortunately seem ill-equipped to do without mathematical assistance. Thankfully, you don't have to be a poker pro like Boeree to benefit from the formula. Bankers and scientists and even some legal scholars rely on the formula, and by the end of this chapter, you'll be able to use it to think more clearly and reason more rationally, too.

A Tale of Two Probabilities

A prerequisite to Bayesian reasoning is viewing life through a probabilistic lens, as we've been doing the past few chapters. But there's a twist here. We'll need to broaden our perspective of probability and grapple with something that has tormented statisticians and scientists for hundreds of years: there are two very different kinds of probabilities—objective and subjective. This can be quite confusing. Math formulas and theorems are all about objective facts; there's no room for subjectivity in math, right? Not so.

The objective form of probability is called *frequentist*, since it measures how frequently things happen. A coin has a 50 percent chance of landing heads and a 50 percent chance of landing tails, because if you flip a coin over and over, you're bound to see heads roughly half the time and tails half the time. Rolling a pair of dice has a 2.78 percent chance of turning up snake eyes (two ones), because that's how often this happens. In very simple settings like these, we don't have to perform any experiments to calculate the frequency of an event; we can deduce the frequentist probability through mathematical reasoning. There are thirty-six pairs of numbers the dice can show—six numbers for the first die, six for the second, so $6 \times 6 = 36$ possible pairs total—each of which is equally likely. Exactly one of these is snake eyes, so its chance of occurring is one in thirty-six, or roughly 2.78 percent.

The problem is that few real-world situations are amenable to this kind of idealized mathematical analysis. What's the probability that intelligent aliens are somewhere out there in the universe? Or the probability that we build AI systems that decide to rebel and enslave all of humanity? Or, more prosaically, what's the probability that candidate X wins election Y? Or that it rains tomorrow? We can't put numbers to these scenarios by counting possibilities the way we can with flipping a coin, rolling dice, spinning a roulette wheel, or shuffling a deck of cards. Nor can we estimate frequencies experimentally, as these are not repeatable events. Our universe either has intelligent aliens or it doesn't. AI either enslaves us or it doesn't. It either rains tomorrow or it doesn't. But one can assign these scenarios numbers indicating how likely one thinks they are to occur. Such *subjective probabilities* quantify the strength of one's belief. Importantly, these beliefs could come from careful mathematical reasoning and data analysis, they could simply be intuitive hunches, or they could be a combination of the two.

The best way to think of subjective probabilities is in terms of betting. If I think Brazil has a 50 percent chance of winning the next World Cup (a subjective probability), that means I'd be willing to place an even-money bet, in which a win earns as much as is wagered. If I give Brazil a 25 percent chance of winning, I'd need my bookie to offer a better payout: a win would have to earn triple my wager. The expected value formula provides the translation between subjective probabilities and betting odds. It's fun to see how this works.

Write p for the probability you believe something will happen, so that $1 - p$ is your probability that it won't, and write w for the amount you wager on it happening and x for the number your bookie multiplies your wager by if you win the bet (in addition to giving your original wager back). If we ignore the vig, which is the bookie's way of adjusting the payout to extract a commission, then a fair bet is one with an expected value of zero: you and your bookie would break even, on average, if your subjective probabilities are accurate. In symbols, this is $xwp - w(1 - p) = 0$, since winning the bet means you earn x times your wager while losing the bet means you lose your wager w. To solve this for x, we can move the losing side of the bet to the other side of the equation, namely $xwp = w(1 - p)$, then divide both sides by w and by p to find that $x = (1 - p) / p$. Plugging in $p = 0.5$ yields $x = 1$, while plugging in $p = 0.25$ yields $x = 3$, as I noted above. But now we have this convenient $(1 - p) / p$ recipe for converting any subjective probability into a wager multiplier.

The most important thing to recognize about subjective probabilities is that they reflect opinion, not fact. And, like an opinion more generally, each person is entitled to one. I can assign whatever probability I want to the possibility of extraterrestrial intelligence, and so can you. One probability being higher than the other doesn't mean one of us is right and the other is wrong. It means one of us finds the possibility more plausible than the other.

Subjective probabilities don't always help us know what's true in life, but they do help us articulate more precisely our views and feelings. After my wife and I had our first child, we weren't sure if we would have another. For a while, we both felt the likelihood (meaning the subjective probability) was around 25 percent. Over time, that number steadily inched upward. After a couple of years, my wife told me she felt about 80 percent in favor of it, and so did I. We decided that was conviction enough to go for it, and we are eternally grateful that we did. (We love you, Peter!)

While the arbitrariness of subjective probabilities takes some getting used to, it's reassuring to recognize that when it comes to situations where the frequentist approach is applicable—things like casino games and repeatable experiments—the two notions of probability generally agree. My subjective probability of rolling snake eyes on a pair of dice is one in thirty-six, because why would it be anything other than the actual likelihood of this occurring? Anyone who thinks the probability of getting a green zero on the spin of a roulette wheel is anything other than one in thirty-seven (or one in thirty-eight for a wheel with a green double zero) either is a fool or knows something about the wheel that I don't know.

Subjective probability vastly expands the number of situations in which we can use probabilistic language and reasoning. And crucially, the individuality inherent in subjective probability is a flexibility we don't get from frequentist probability. This flexibility underpins the ability of Bayes's formula to help each of us think more clearly about the issues we face in our lives.

The 250-Year-Old Formula

Thomas Bayes is known to have published only two works in his lifetime. The first, called *Divine Benevolence*, addresses a matter of religious interest: "Unless we know what God is, and entertain clear conceptions of his perfections, and particularly of his moral attributes, we shall neither know how to behave towards him, nor what we are to expect from him according as we do behave." The second work of Bayes is quite a different matter. It's a defense of Isaac Newton's foundations for calculus.

Bayes held a lifelong interest in mathematics, and while he was an amateur in the subject, he left an indelible mark in the discipline with his eponymous formula. Curiously, Bayes never published the famous formula that bears his name. It was discovered posthumously among his piles of notes and writings by his friend Richard Price.

Price was a brilliant Welsh minister and political activist who championed liberal causes, including the American and French Revolutions. He corresponded with the Marquis de Condorcet, whom we met in chapter 3 when discussing Condorcet's jury theorem, as well as notable American political figures of the day, including George Washington, John Adams, and Thomas Jefferson. Like Bayes, Price was keenly interested in both religion and mathematics. This combination of interests, and a desire to wed them, is why the formula he found in Bayes's belongings first caught his eye. Price felt that Bayes's formula helped prove the existence of God.

Price revised and expanded Bayes's unpublished paper containing the formula, and while doing so he added an extended introduction. In it, he explained (albeit in rather rambling prose) why he felt Bayes's work was evidence of divine intervention: "The purpose I mean is, to show what reason we have for believing that there are in the constitution of things fixt laws according to which things happen, and that, therefore, the frame of the world must be the effect of the wisdom and power of an intelligent cause; and thus to confirm the argument taken from final causes for the existence of the Deity." In essence, he found Bayes's formula such a shockingly powerful way of taming randomness that he viewed it as a sign of what today we would call intelligent design.

This theological interpretation of Bayes's formula has not stood the test of time. But the formula itself has risen over the years to become one of the most important in all of mathematics. As Price wisely discerned, it brings mathematical precision to bear on matters that would otherwise be relegated to the realm of pure guesswork.

Bayes's formula can help with calculations involving frequentist probability. But it's with subjective probability that the formula really shines. That's because, for subjective probability, Bayes's formula provides a recipe for revising the strength of your convictions in light of new evidence. In doing so, it provides a numerical framework for an important task normally done through instinct alone.

The only math concept we need to discuss before stating the formula is *conditional probability*. This is the likelihood of something given that something else has already occurred or that some condition holds. It allows us to squeeze context into our calculations. If you randomly select a person in the US, the probability that they experience a white Christmas next year (at least an inch of snow on December 25) is 3.5 percent. But the probability of this given that the person lives in Alaska is 66 percent. The probability that a person's death will result from heart disease is about 20 percent. For men, this figure rises by five percentage points—meaning the probability of dying from heart disease given that you're a man is 25 percent.

In mathy notation, we write the probability of a thing X happening as $P(X)$, and if Y is the context we want to add, we write the conditional probability of X happening given Y as $P(X|Y)$. One curious property of conditional probability is that it matters which is the event of interest and which is the context, meaning $P(X|Y)$, the probability of the event given the context, is usually not equal to $P(Y|X)$, the probability of the context given the event. The probability of a white Christmas given that you chose someone in Alaska is 66 percent, but the probability that you chose someone in Alaska given that they experienced a white Christmas is only 4 percent. Alaska doesn't have many residents, so the probability of choosing someone there is small despite the additional context of experiencing a white Christmas.

Bayes's formula pins down the precise relationship between the two kinds of conditional probabilities: the event given the context and the context given the event.

$$P(X|Y) = P(Y|X) \times \frac{P(X)}{P(Y)}$$

In words, to flip a conditional probability around, just multiply by the probability of the event and divide by the probability of the context. That's it. That's the formula. It doesn't look like much, does it? Honestly, it's not. It takes only a couple of lines to derive this formula from first principles and officially establish it as a mathematical fact. The power of the formula comes from how we interpret it.

The first step is to think of Y not as additional context but as a new development that might cause us to rethink the event X. For the probability of intelligent aliens, this might be leaked government documents describing unidentified flying objects. For the probability of AI enslaving us, the new development might be reports of the latest version of ChatGPT manipulating users in dangerous and deceptive ways. For the probability of a candidate winning an upcoming election, it might be a new poll that comes out, a debate that takes place, or an economic event that throws a wrench in the race. These new developments need not be so compelling that they cause us to reverse course and think that something we previously thought improbable is now probable. These developments can be anything that could conceivably impact the probability (objective or subjective) of the event we're considering.

The second step to interpreting Bayes's formula is a simple algebraic one. Rather than highlighting the flipped conditional probability $P(Y|X)$ out front the way it's written above, let's relegate that to the fraction in the formula and instead bring out $P(X)$ as the term of interest:

$$P(X|Y) = P(X) \times \frac{P(Y|X)}{P(Y)}$$

This looks pretty darn similar to the earlier version, and the two are mathematically equivalent. But now something tremendously important leaps out: this formula shows us how to update the probability of an event X in light of the new development Y. To do so, multiply the probability of the event by the fraction appearing in this second version of the formula.

Suppose there's a pandemic—for lack of imagination, call it COVID 2.0—and that 10 percent of the people in your region with flu symptoms have this new virus. Alas, you develop flu symptoms. You take a COVID 2.0 test and it comes up positive. But that's not a guarantee you have the virus, because tests can produce false positives. You look up the false-positive rate and find it is 5 percent, which means that 95 percent of all positives correctly indicate the presence of infection. Now you're really confused. The prevalence of COVID 2.0 among people like you with flu symptoms suggests that you have a 10 percent chance of infection, while the positive test result and the false-positive rate suggest that you have a 95 percent chance of infection. So, which is it: 10 percent, 95 percent, or something in between?

We can think of the 10 percent rate as our initial estimate of the probability of infection, an estimate that needs to be updated after the news of your positive test result. In other words, this is a matter of updating a probability in light of a new development—precisely the purview of Bayes's formula. Before seeing how to do the math and what the answer is, try taking a guess: What do you think your likelihood of infection is in this situation?

Now let's give the formula a shot. In what follows, we'll focus on the people in your region with flu symptoms. We need to compute the probability of having COVID 2.0 given that you tested positive. Bayes's formula says to start with your initial probability of infection, 10 percent, then update it by multiplying by the probability of a positive test given infection and dividing by the overall probability of a positive test. So we need to figure out those two numbers. The probability of a positive test given infection is called the *sensitivity* of the test. You can usually look this up online or ask your pharmacist or doctor about it. The at-home antigen tests for COVID-19 had a sensitivity of 80 percent, so let's just use that same number here for our hypothetical COVID 2.0. Figuring out the overall probability of getting a positive test result is a little trickier, but we can do it. We just need to add up the two different ways of getting a positive test result: true positives and false positives.

True positives are people who have COVID 2.0 and whose

test correctly detects it. Since 10 percent of our population has COVID 2.0 and 80 percent of these cases will correctly trigger a positive result (due to the test's 80 percent sensitivity), 8 percent of our population are true positives. False positives are the people in the 90 percent of the population without COVID 2.0 whose test falsely comes back positive. Since the test has a 5 percent false-positive rate, our population has $0.9 \times 0.05 = 4.5$ percent false positives. Adding the true positives and false positives shows that the overall likelihood of a positive test in our population is 12.5 percent. Bayes's formula then tells us that our original 10 percent probability should be updated to $0.1 \times 0.8 / 0.125 = 64$ percent.

How did you do? I suspect most readers guessed a higher probability, something closer to the 95 percent we'd get by considering only the test's error rate. If you don't believe me that even with a positive test result your chances of infection are only 64 percent, think about this. If only one in a million people has the virus, then a million people taking the test would yield fifty thousand positives, even though only one of these fifty thousand would be a true infection. In that case, even if you test positive, you have only a 0.002 percent chance of infection. I hope you keep this in mind if you or a loved one is pregnant and screening the baby for rare disorders. Positive test results in situations like that are not nearly as scary once you properly take into account the prevalence of the disorders being screened for.

The issue here is that our brains are unfortunately terrible at combining different pieces of probabilistic information. We usually stick to our original estimate and largely ignore any new evidence, or we overcompensate and let the new evidence dominate our thinking rather than carefully layering it on top of our earlier thoughts. In short, we tend to weigh new evidence either too heavily or too lightly. Bayes's formula helps us achieve the right balance, although it admittedly takes some practice to use it effectively.

To see how easy it is for intelligent, educated people to botch probabilistic considerations when they don't rely on Bayes's formula, let's step back to the 1990s. Specifically, to the infamous OJ Simpson trial. Alan Dershowitz, part of the legal defense's "dream team," as it was called, wrote a bestselling book about the trial shortly after it concluded. In it, he writes, "We knew that we could prove, if we had to, that an infinitesimal percentage—certainly fewer than 1 out of 2,500—of men who slap or beat their domestic partners go on to murder them." This statistic formed the basis for his argument that domestic abuse should be inadmissible evidence in a murder trial in which a man who is known to have beaten his wife is accused of later killing her. But in a brief letter published the next year in *Nature*, a statistics professor at Virginia Tech, Jack Good, said not so fast, Mr. Dershowitz.

Good notes that the low frequency of domestic abusers who go on to be domestic murderers is not the relevant statistic. Or rather, it's not the whole story. He says that by the same reasoning, it would be inadmissible to mention that the defendant was married to the murder victim, since very few husbands murder their wives. I'll spare you the detailed calculation, but in short Good uses Bayes's formula to show that, absent any other information, if a woman is murdered and her husband physically abused her previously, then there's a 90 percent chance he is the murderer. One can't use this figure alone to convict OJ Simpson or any other defendant, because the goal of a trial is to surface information specific to a particular case. But this shows that based on aggregate numbers alone, Dershowitz had it backward. An abusive husband is not an exceedingly improbable culprit when his wife is murdered. He is an exceedingly probable one.

Jack Good had an illustrious career that I can't pass by without mentioning. His interest in Bayes's formula dated back to World War II. He worked with Alan Turing to crack Germany's Enigma code, a story told to popular audiences in the 2014 movie *The Imitation Game*. What you might not know from the film is that Good and Turing used a statistical approach to crack the code that relied heavily on Bayes's formula. After the war, the two continued to collaborate, working on designs for computers and on advances in Bayesian statistics. Years later, Good served as an expert consultant to Stanley Kubrick for the movie *2001: A Space Odyssey*.

Bayes's formula is widely used today in highly quantitative disciplines. When you get in a car accident, your insurance rate jumps up. Did your insurance company simply guess how likely a subsequent accident is? Nope. They almost surely used Bayes's formula to update the probability that you'll be in an accident after your recent one. When finance professionals listen to remarks by the chair of the Federal Reserve bank, they use Bayes's formula to update their probability of interest rate adjustments.

Bayes's formula even helps email apps filter spam and self-driving cars make decisions on the road.

It's not easy bringing Bayes out of these professional, number-centric realms and into the hands of ordinary people to help navigate everyday situations. But I'm going to try. I think the key is recognizing that when we're using Bayes's formula, we're often dealing with subjective probabilities. And subjective probabilities are a bit like the TV show *Whose Line Is It Anyway?*, "where everything's made up and the points don't matter." That's a bit of an overstatement, but my point is that subjective probabilities don't have to be carefully calculated numbers like the ones in the above examples.

My advice is to try coming up with ballpark estimates for the subjective probabilities you need, then use Bayes's formula to see how these estimates should be updated in light of new developments. This approach offers the flexibility of intuitive thinking while reinforcing it with mathematical scaffolding to keep your intuition on track. To help you do this, I'll present a new, never-before-published recipe that makes using Bayes's formula as easy as ABC. Well, technically it's as easy as $A - B + C$, but you'll need to read the next section to see what that means.

Bayes Made Easy

Instead of working with actual probabilities, let's use numbers that represent some common ballpark probability estimates, with bigger numbers indicating rarer events:

0 is almost certain (close to 100 percent)

1 is one in two (50 percent)

2 is one in four (25 percent)

3 is one in ten (10 percent)

6 is one in a hundred (1 percent)

10 is one in a thousand (0.1 percent)

20 is one in a million (0.0001 percent)

First, choose the number on this list that best expresses your subjective probability of the event or belief you are interested in; call this A. Next, choose the number B that you think best represents the overall probability of the development or context that you want to take into account. Then, choose a number C for the probability of the development or context given the event or belief. Finally, plug these numbers into the formula $A - B + C$, then use the list above to convert the number you get back to a probability. This is your revised subjective probability for the event or belief in light of the new development or context.*

---

*If you're wondering what's going on and why this works, here's the explanation. It would be nice to use powers of ten, like Romer did when pricing risk, for all the probabilities in Bayes's formula. But powers of ten are too spread out for commonly occurring probabilities. We'd have near certainty, one in ten, one in a hundred, one in a thousand, and so on, with nothing between these values. Instead, I'm using powers of two here. The number in the list indicates approximately what value x gives the corresponding probability 2^{-x}, which if you recall negative exponents is another way of writing $\frac{1}{2}^x$. I rounded the percentages a bit to make them

If the number you get in the last step when calculating $A - B + C$ isn't on the list, just ballpark a probability by looking at the nearby numbers. For instance, if you get five, then your probability is somewhere between 10 percent and 1 percent but closer to 1 percent, so you could call it, say, 3 percent. When making your initial probability estimates, and this conversion at the end, just remember our motto from chapter 5: close enough for jazz. All you're doing here is shoehorning ballpark numbers into Bayes's formula to be a bit more accurate about updating beliefs than you would be without any math.

Imagine that a friend tells you about a hot investment opportunity you gotta jump on, but you're not sure if you should trust her with your hard-earned savings. You're feeling about 50/50. To help clarify matters, you decide to do a test. You show her a list of ten stocks and ask her to pick the one she thinks will rise the most in a month. Sure enough, she nails it and picks the winner. This certainly increases the amount you trust her on financial matters, but by how much?

Since your initial confidence in her abilities is 50 percent, you choose $A = 1$. Overall, there's a one-in-ten chance of picking the winning stock on your list of ten, so you select $B = 3$. It's harder to estimate the probability that she picks the winning stock given that she's good at investing, but let's give it a try. A

---

easier to work with. So 1 is $\frac{1}{2}$, or 50 percent; 2 is $\frac{1}{4}$, or 25 percent; 3 is $\frac{1}{8}$, or 12.5 percent, which I'm rounding to 10 percent, and so forth. If you write all the probabilities in Bayes's formula as negative powers of two like this, you get $P(X|Y) = P(X) \times P(Y|X)/P(Y) = 2^{-A} \times 2^{-C} / 2^{-B} = 2^{-(A-B+C)}$, so the number $A - B + C$ represents your revised probability of event or belief X in light of development or context Y.

financial expert has a much better than random chance of picking the best-performing stock on a list, so we need a probability bigger than 10 percent. But no matter how much expertise someone has, there's still a lot of uncertainty and luck involved in picking stocks. I doubt even an expert could win this stock-guessing game more than half the time, so let's go with a probability that's less than 50 percent. That leaves only one choice: 25 percent, hence $C = 2$. We then plug in: $A - B + C = 1 - 3 + 2 = 0$, which translates to 100 percent. We shouldn't interpret this as a guarantee that the friend knows her stuff, since we're just ballparking odds here. But it shows that by passing our stock-guessing game with flying colors, our confidence in her should rise dramatically and allay our initial doubts.

Sometimes when doing calculations like this you'll end up with a negative number. If so, just round it up to zero and interpret it as "exceedingly likely." You can use this Bayes-made-easy procedure for factual matters concerning frequentist probabilities, but in my experience it can be tricky to get all the probabilities you need, even with ballpark estimates. I suggest using it primarily for subjective probabilities where you just want to gauge how much more or less confident you should be in something after learning something new about it.

Bayesian Reasoning

Even if you don't plug numbers into Bayes's formula and use it quantitatively, there are important qualitative lessons on thinking clearly to distill from the formula and our interpretation of it. The first is that the experiences we have and the information we encounter are filtered through our prior experiences and knowledge. The formula shows that our prior beliefs are updated, not replaced, as we proceed through life. I suspect that this plays a large role in many disagreements and divisions in society.

In March 2024, I watched then-president Biden's State of the Union address and was amazed by how differently it struck me and some of my acquaintances. It left me thinking, How could we watch the same speech and come away with such different reactions? I was writing this chapter at the time and found that thinking about Bayes's formula helped me come to terms with the situation. I watched that speech with an already formed opinion of the president, the economy, the wars in Ukraine and Gaza, and so on. What I saw that evening caused me to update these opinions, not to form new ones. When I judged the speech afterward, I think it was less that I was judging the speech itself and more that I was judging the impact it had on my prior beliefs. And same for my acquaintances. We all watched the same speech, but because we went in with different prior beliefs, the speech had different impacts on us.

You might object and say, Wait a minute, the point in this chapter is to encourage us to use Bayes's formula in our thought processes, so why are you talking as though we already do? The answer is that I think Bayes's formula conveys the general method by which our brains process information. It's just that we do so without the benefit of numbers and the mathematical rigor of the formula, so we often update our beliefs too much or too little.

In other words, I believe that in broad strokes we process information and update our beliefs in a somewhat Bayesian manner, just not as accurately as we would with the aid of the formula.

I can't prove that this is how our minds work, and there is debate about this in the scientific literature. Consider it simply my opinion on how we think. I'm willing to change this opinion in light of new evidence, but for now I put a high subjective probability on it. And so does Liv Boeree. When asked whether she uses Bayes's formula while playing poker, she answered, "Do I do it mathematically in game? Absolutely not. I don't have likelihood ratios and so on in my head. But ultimately our brains are Bayesian machines."

A second lesson is that we tend to believe rare events are meaningful on the basis of their rarity alone, but that's a mistake. Bayes's formula shows that we shouldn't focus on an event in isolation. We should focus instead on how the event impacts our beliefs. And for assessing the impact of an event on one of our beliefs, what matters is how much more or less rare the event would be if our belief were valid. That's because the formula says to update our belief in X by multiplying it by the ratio $P(Y|X) / P(Y)$, the rarity of an event Y assuming X is valid relative to the overall rarity of Y. Let's unpack this lesson with a couple of examples.

If someone claims to have won the lottery, I shouldn't assume they are lying just because the jackpot is a one-in-a-million event. But if I first suspect that someone is a fraud and *then* they claim to have won the lottery, the improbability of winning lends tremendous credence to my initial suspicion. Indeed, the probability of claiming a win given that they're a fraud is much greater than the overall probability of claiming a lottery victory—so I should significantly increase my subjective probability that they're a fraud upon hearing their claim of winning the lottery.

If the stock market plummets, that must mean the president did a terrible job handling the economy, right? Not necessarily. That's falling into the trap of viewing a rare event in isolation rather than assessing its impact on a prior belief. You should first form your opinion of the president's handling of the economy based on everything else you've seen, things like bills the administration supported, executive orders it signed, and interest rate changes if the president appointed a new chair of the Fed. The market crash is just one data point among many.

In complex situations like this, I usually find that it's too challenging to ballpark the numbers Bayes's formula needs. But I still find the formula helpful for pointing my mind in the right direction. Suppose your prior estimate is that the president has been quite competent in economic matters. To update this estimation in light of the market crash, Bayes's formula says you'd need to ballpark how much an economically competent president would reduce the likelihood of such a crash. This is a hard question to answer, but it's the right one to ask. It's essentially the question of whether the crash was the fault of the president or instead due to matters beyond the president's control. This is the relevant question for determining whether you should reassess your opinion of the president's economic competence.

This highlights a third lesson: even when Bayes's formula doesn't provide the answers we seek, it guides us to the right questions to ask.

Back to Anxiety

A close family member, who I'll call Laura, recently had her first mammogram. She was nervous about the experience, and that feeling grew into unbridled panic when the doctor's office called. "There's an anomaly," they told her. "It's probably harmless, but it could be cancer; a follow-up exam is needed." The days waiting for that exam were excruciatingly long and fretful. "Probably harmless" sounds good until you're the one facing those words. Laura soon did what any of us would do in that situation: she googled the condition to find the likelihood that it was cancer. The odds were indeed low, around 5 percent, but a one-in-twenty chance of the dreaded C-word was not low enough to avoid a full-blown panic attack.

After a couple more days of suffering in anxious uncertainty, Laura decided to see what a Bayesian perspective would reveal. (Conveniently, she had just proofread a draft of this chapter and was happy to give the material a try.) It turns out that a woman of her age and family history has an extremely low likelihood of breast cancer, *much* lower than 5 percent. Rather than the mammogram anomaly meaning that she suddenly jumps up to a 5 percent chance of cancer, the anomaly should be seen as new evidence for which her prior chance of breast cancer should be adjusted. Since her prior likelihood was so low, even a modest increase meant it was still very unlikely. And she learned that several of her family members had similar anomalies on their first mammograms, all of which turned out to be benign, further lowering the weight of her mammogram news.

Without plugging a single number into Bayes's formula, Laura suddenly felt tremendously relieved. She didn't need a new figure to replace that 5 percent with, one that she'd surely obsess over no matter how small it was. All she needed was the Bayesian realization that she's only a tiny bit more likely to have cancer now than she was before the mammogram news—and if she wasn't too worried about cancer before, it wasn't rational to suddenly become so terrified of it. Thankfully, the Reverend Bayes did not lead her astray: the follow-up exam arrived and delivered the good news we were all hoping for.

7

The Secret Social Media Formula

In December 2022, one of the most viral videos on TikTok featured a young woman posing in front of a bathroom mirror. With one hand, she holds up her phone. With the other, she runs her fingers through her hair. This banal scene is overlaid with the following text: "Imagine how good your life would be if you had a 26yo nursing assistant by your side, now replace S with N." That's it. That's all that happens.

The internet was flummoxed. Comments rolled in, most simply expressing puzzlement. "Someone please tell me what a nurning anninntant is, I'm loning my mind here," wrote one viewer. "I watched this both under the influence and sober and it still doesn't make a shred of sense," wrote another. One user shared it on Twitter, racking up thirty million views and a quarter-million Likes, with the remark "Every day, at least once I am haunted by this." He continued: "This, for me, remain[s] the biggest unsolved intellectual question of our era."

Others offered far-fetched explanations, none of which were convincing. "She's single, so she's Ms. Registered Nurses are called RN. If she was by your side, she'd be Mrs. If you replace the S, MRS -> MRN, Medical Registered Nurse (she doesn't fully understand the title/letter system)," wrote one viewer who had clearly spent too much time thinking about it. Another took it in a more philosophical direction: "I think it's like replace South with North. Like turn your whole perspective around. Now be someone's nursing assistant because you've realized how great it will make their life." Yet another viewer commented, "Even AI can't solve it," attaching a screenshot of ChatGPT's bungled attempt to decipher the conundrum.

After weeks of rampant speculation and frustration, the mystery was resolved—sort of. When *Daily Dot*, an internet culture news site, messaged the content creator asking her to explain the meaning of her viral riddle, she replied, "It's meant to be a brain stump. It doesn't mean anything." But this response raises more questions than it answers. Why would she post such a pointless video, and why did TikTok's algorithm show it to millions of users?

The answer to the first of these is easy. The TikToker posted it because she suspected it would generate a lot of traffic, some of which might be drawn to the OnlyFans account linked in her profile. As for the second question—why did TikTok push this exasperatingly nonsensical time-trap to millions of users?—the answer is that it wasn't just the viewers who were duped. The algorithm was, too.

To understand how, we'll need to come to grips with a math formula that undergirds not just TikTok but nearly all social media platforms. This formula explains the viral success of the nursing assistant video, along with so much else of what you see online. The more you know about this formula, the more you'll be able to take charge of your social media experience and fill your feed with the posts you enjoy the most instead of crap that's a clickbait trap.

This formula first escaped the clutches of the tech giants a few years ago at Facebook. Let's start with that remarkable story before turning back to TikTok.

A Series of Scandals at Facebook

A 2014 research paper published in the *Proceedings of the National Academy of Sciences* found "experimental evidence of massive-scale emotional contagion through social networks." That may sound like typical academic jargon, but this was not a typical study. The lead author was a data scientist employed by Facebook, and the test subjects in this experiment were, well, *us*. Across a one-week period, Facebook adjusted the feeds of over half a million users, showing some people more positive content than usual and others more unhappy content. The question was whether this would impact their behavior online, and the answer was yes. The revelation that people's emotions were toyed with by a tech giant, without their consent or even their awareness, caused an uproar. *The Atlantic* ran a story provocatively headlined "Everything We Know About Facebook's Secret Mood-Manipulation Experiment." The article opens with the line "It was probably legal. But was it ethical?"

The brouhaha over this study brought to the fore the dubious morality of subjecting billions of users to the whims of a company willing and able to experiment on them. And it highlighted just how valuable Facebook's vast troves of user data are. As we'll discuss in chapter 8, the value of the information Facebook collects on all of us is not merely scientific; it also underlies the company's financial success. It's no surprise that others are interested in gaining access to this prized cache. But it was quite a shock when it emerged in 2018 that a company, Cambridge Analytica, had pilfered personal data from eighty-seven million Facebook users and used it to support the 2016 campaigns of both Trump and Brexit.

These were a couple of the more notable Facebook scandals in the 2010s, but they were not the only ones. Distrust over Facebook's handling of our emotions and our personal data rose throughout the decade, as did demands for more transparency from the company. This public consternation would not abate in the next decade. In 2020, Netflix released the docudrama *The Social Dilemma*, which sought to expose the dangers of social media. Later the same year, *The Atlantic*'s executive editor, Adrienne LaFrance, penned a widely shared article titled "Facebook Is a Doomsday Machine."

In response to these salvos from print and streaming media,

Facebook's Nick Clegg* wrote a lengthy blog post titled "You and the Algorithm: It Takes Two to Tango" in which he practically throws Facebook users under the bus. The algorithm he's referring to is the one that decides which posts users see, and the gist of his argument is that since this algorithm is shaped by what users do on the platform—as we'll discuss momentarily—it's their own fault if Facebook causes any harm: "The personalized 'world' of your News Feed is shaped heavily by your choices and actions. . . . Perhaps it is time to acknowledge it is not simply the fault of faceless machines? . . . We need to look at ourselves in the mirror, and not wrap ourselves in the false comfort that we have simply been manipulated by machines all along."

While primarily aiming to deflect the accusations against his company, Clegg's five-thousand-word screed does raise some valid points about transparency and user empowerment. "Faced with opaque systems operated by wealthy global companies," he writes, "it is hardly surprising that many assume the lack of transparency exists to serve the interests of technology elites and not users. In the long run, people are only going to feel comfortable with these algorithmic systems if they have more visibility into how they work and then have the ability to exercise more informed control over them." Later in the post, he writes: "You should be able to better understand how the ranking algorithms

---

*If the name sounds familiar, that's because Clegg was a career politician in the UK who served as the deputy prime minister from 2010 to 2015. In 2018, he accepted Mark Zuckerberg's offer to become Facebook's vice president for global affairs and communications, and he has since been promoted to president for global affairs at Meta, the parent company of Facebook, Instagram, and WhatsApp.

work and why they make particular decisions, and you should have more control over the content that is shown to you."

However, it's one thing to say there should be more transparency; it's another to provide it. Clegg's rambling manifesto offers disappointingly little technical insight, in my opinion. That said, two months before Clegg published his piece, a few Facebook employees posted an article on the company's engineering blog, tagged under the exciting categories of "core infra" and "machine learning applications," that appears to offer much of what Clegg's post doesn't. This article even includes a fancy-looking formula central to Facebook's elusive algorithm. I'm a math professor who writes about social media algorithms, yet to this day I struggle to decipher what they wrote. The problem is they didn't bother to explain half the symbols in the formula, so all we can do is guess based on their cryptic labels. I'm not sure if Facebook published this formula to counter the "lack of transparency" Clegg decried or merely to give the superficial impression of doing so.

Let's zoom out for a second. When a user logs on to Facebook, there are thousands of posts they could be shown: all the recent posts from their friends and from the pages and groups they follow, as well as some older posts that have had recent activity. An algorithm is used to determine the order in which these posts show up in the user's feed. Most users scroll through only the first handful of posts, so the ranking that the algorithm comes up with matters a lot. To understand why some posts go viral while others wallow in obscurity, we must understand this algorithm. There are other algorithms at work at Facebook, such as the one that recommends friends and pages to follow, and the search algorithm powering the search bar. But when people speak of *the* Facebook algorithm, it's the post-ranking system they have in mind.

The engineering blog post does a decent job of presenting some technical aspects of the algorithm, but the key to understanding how posts are ranked is the formula that the authors leave conspicuously unexplained. When I first read that blog post and encountered the formula in it, I felt like the cave dwellers in Plato's famous allegory seeing shadows dance upon the wall—flat, colorless projections of a richer world existing out of sight. I knew the obfuscated formula in this blog post was not the full story. I just didn't know how to get outside the cave and find the real math behind it. Half a year later I got a phone call from Will Oremus, a technology writer at *The Washington Post*, that marked the beginning of the end of the mystery.

In September 2021, Facebook was rocked by one of the biggest scandals ever to strike the tech industry. Frances Haugen, a Facebook product manager turned whistleblower, snuck over ten thousand pages of corporate documents and internal messages out of Facebook headquarters. She did so by painstakingly taking pictures on her cell phone of each page and screenful of information, one at a time to avoid triggering the security system that goes off when employees download sensitive documents. She leaked these to a handful of press organizations, *The Washington Post* among them.

The news cycle was soon dominated by this event. For weeks, the media outlets with access to Haugen's cache ran article after article detailing what was uncovered. As one would expect, these generally focused on the most alarming, attention-grabbing revelations. Internal studies documented Instagram's harmful impact on the mental health of vulnerable teen girls. A secret whitelist program exempted VIP users from the moderation system the rest of us face. Mark Zuckerberg and other executives were allegedly unwilling to stem the floods of dangerous extremist content propagated on the platform if doing so would reduce the amount of time users spent glued to their feeds.

Will Oremus, the journalist I spoke with one breezy September afternoon amid this media maelstrom, had something else in his sights. He wanted to lift the veil on the formula in the engineering blog post, and he wanted my help in doing so. It turns out the leaked documents he had seen tell a fascinating story— one that brings us back to the weighted sums of chapter 2 and the expected value formula of chapter 4.

The Facebook Formula

Haugen's documents reveal that Facebook engineers have assigned a point value to each type of engagement users can perform on a post. These engagements include clicking the various reaction buttons (such as the Like button and the angry face emoji), leaving a comment, and resharing the post. For each post you could be shown, these point values are multiplied by the probability that the algorithm thinks you'll perform the corresponding form of engagement. These multiplied pairs of numbers are then added up, and the resulting total is the post's personalized score for you. There's a bit more to it than this, such as adjustments to ensure that you don't see too many posts from the same user or too many videos in a row, but in broad strokes your feed is created by sorting posts according to these scores, from highest to lowest.

The formula I just described for computing these personalized scores is none other than the expected value formula. This is a little easier to see with symbols. Suppose we have a specific user and a specific post in mind, and we write P_{Like} for the probability that the user Likes the post, P_{Love} for the probability that they tap the heart emoji, P_{angry} for the probability that they tap the angry emoji, $P_{comment}$ for the probability that they comment on the post, and P_{share} for the probability that they share it. (There are other forms of engagement, but let's just focus on these for now.) And let's write V_{Like}, V_{Love}, and so on for the point values assigned to these engagements. Then the magic formula is:

$$\text{Score} = V_{Like} \times P_{Like} + V_{Love} \times P_{Love} + V_{angry} \times P_{angry}$$
$$+ V_{comment} \times P_{comment} + V_{share} \times P_{share}$$

This is the post's expected value for you: the number of engagement points, on average, that the algorithm expects you to provide on the post. The "on average" here is a bit misleading, because if you show someone the same post multiple times, their probabilities of engagement will not stay the same. If I see a picture of my cousin's glistening new engagement ring, I'll likely tap the heart emoji and write a nice comment. But if I see the exact same picture ten times in a row, don't count on me clicking anything by the tenth one.

It might be better to forget the word *average* and just view this expected value as a weighted sum. The idea is that the algorithm wants to surface the posts you're most likely to engage with—but there are several forms of engagement, not just one. It wouldn't make sense to treat all forms of engagement equally; a reshare really does seem like stronger engagement than a Like. So the different forms of engagement are weighted differently, and a weighted sum combines them into an overall measure of anticipated engagement.

Let's try this with some concrete numbers. Suppose a Like is worth one point, a heart emoji is worth five points, and a comment is worth thirty points. And suppose one of your friends posts a picture of the puppy they just adopted, while another friend writes a post about a new job they landed. You're fond of both friends, but let's be real: you're more excited about the puppy than the job. If there's a 50 percent chance you'll Like the puppy pic, a 20 percent chance you'll Love it, and a 10 percent chance you'll comment on it, then the puppy post scores $1 \times 0.5 + 5 \times 0.2 + 30 \times 0.1 = 4.5$. If there's a 20 percent chance you'll Like the job announcement post, a 10 percent chance you'll Love it, and a 5 percent chance you'll comment on it, then its score is $1 \times 0.2 + 5 \times 0.1 + 30 \times 0.05 = 2.2$. Since $4.5 > 2.2$, the puppy pic wins and is placed higher in your feed than the job announcement.

Now suppose there's also a post by your uncle claiming that COVID was caused by 5G towers. You're unlikely to Like or Love this post; let's go ahead and give both of those 0 percent chances of happening. But you are tempted to write a comment telling your uncle he's full of shit, or at least politely explaining why he's wrong. Let's put your probability of commenting on this post at, say, 20 percent. Then its score is $1 \times 0 + 5 \times 0 + 30 \times 0.2 = 6$. Since $6 > 4.5 > 2.2$, before you come to the puppy post that makes you happy and the job post that makes you mildly envious, you're going to see a COVID conspiracy post that boils your blood. Facebook doesn't *try* to make you angry, but the algorithm has figured out what kinds of posts will keep you engaged—and unfortunately, it's not always the ones you want to see.

How do you take charge and tame the algorithm? The *V*s in the formula, the engagement point values, are out of your control. But you can influence the *P*s, the estimates of your engagement probabilities. If you tend to engage with posts about food, then over time the algorithm will bump up your estimated engagement probabilities on food posts. If you want more food content in your feed, go ahead and Like, Love, comment, and share away. If you don't want food content, don't engage with it.

Where things get subtle is with unpleasant posts, especially ones that anger or offend you. You don't want to engage with this stuff, thereby training the algorithm to show you more of it. But you also don't want to give the poster a free pass and scroll by without pushing back on the bad content. While Facebook has never disclosed any of the engagement point scores, Haugen's leaked documents provide the missing piece of the puzzle. They reveal that the angry emoji is worth zero points, so it's a convenient way to tell the poster how you feel without also telling the algorithm you want more.

Think of it this way. When you argue with your uncle, you are giving his COVID conspiracy posts thirty points for each comment you leave—no matter how critical your comment is—and these points drive all his other COVID conspiracy posts up in your feed. In fact, these points drive his COVID conspiracy posts up in *everyone's* feed, because the algorithm is smart enough to realize that if you're inclined to comment on these posts, so are his other friends. Worse still, the algorithm associates COVID conspiracy content with other conspiratorial content, so when your uncle posts a flat-Earth link, the algorithm in essence thinks: "They commented on the other conspiracy posts, so I bet they'll comment on this one, too."

It doesn't end there. The algorithm correctly deduces that if you're likely to comment on your uncle's conspiratorial content, you're likely to comment on other users' conspiratorial content as well. In the end, your laudable effort to educate your uncle with a choicely worded comment backfires and tells the algorithm to elevate all conspiratorial content in your feed and, to a lesser but non-negligible extent, in the feeds of other users. Oops.

You might be thinking, "Why can't Facebook's brilliant AI figure out which comments are positive and which are negative, then give positive points for positive comments and negative points for negative comments?" Part of the answer is that it's harder than you might expect for AI to decode language like this. But the bigger issue, I suspect, is that there's simply no motivation for the company to rejigger its engagement valuation this way. Every minute on the platform is a minute of ad revenue, and it takes just as long to type a negative comment as it does a positive one.

After seeing how engagements are scored on Facebook, you'd be wise to resist the urge to comment on your uncle's conspiratorial rants and simply leave an angry emoji instead.

Interestingly, the angry emoji hasn't always been worth zero points. When Facebook first introduced emoji reactions in 2016, adding to the original Like button, their point values were all set to five. Over the next four years, employees repeatedly voiced concerns that angry reactions occur disproportionately on toxic posts. There was data collected to support this hypothesis. There were internal proposals to lower the point value of angry reactions, aiming to reduce the virality of bad stuff. And there was a sequence of (at times begrudging) acquiescences by management to lower the point value, first from 5 to 4 and then from 4 to 1.5. In September 2020, the point value was finally dropped all the way to zero.

Facebook recently revealed that other forms of engagement in the weighted-sum formula include: clicking to show more comments on a post, clicking on a photo in a post, watching a video in a post, RSVPing to an event in a post, following the page that shared a post (for posts from pages rather than groups or friends), sending a post to someone in a message, and spending time viewing a post instead of scrolling by right away. But it hasn't revealed the point values for these engagements or any others. I find this a half-hearted nod toward Nick Clegg's talk of transparency. And I mean that literally. Half the formula is engagement types, the other half is engagement values—and while Facebook has finally told us half the story (after much of it leaked anyway), Facebook continues to hide the other half from us.

Most forms of engagement elevate posts in your feed, but there is one that does the opposite. Reporting a post for a policy violation is a factor in the engagement formula, but it's one with a negative weight. That means the more likely the algorithm thinks you'll report a particular post, the lower the post appears in your feed. I'm not suggesting you *should* use this technique to demote posts about the Kardashians or SoulCycle. I'm just saying you *could* do so if you wish.

In the discussion so far, I've treated the Vs in the engagement formula as fixed numbers indicating how heavily each form of engagement is weighted. The Vs do this, but that's not all they do. They also specify how much the *topic* of the post is weighted. If Facebook executives want users to see less political content, they can pull an algorithmic lever that lowers the Vs for all political posts. And the Vs account for the relationship between the user who sees the post and the user who posted it. Facebook can make engagement from family and close friends score higher, or it can go in the opposite direction and give more points to engagement from strangers. These adjustments to the Vs are out of your sight and out of your control. But the bottom line is that you are the master of the Ps, and those matter in the formula just as much as the Vs.

Nick Clegg's "It Takes Two to Tango" blog post includes a

section titled "How to Train Your Algorithm" that purports to tell users how to take responsibility for their feeds by choosing their engagements thoughtfully. I wish Facebook leadership made an honest effort to explain the engagement formula, rather than leaving it to outsiders to piece together from confusing blog posts and leaked company documents. And I wish they revealed the engagement point values, not just the engagement types, since those tell us exactly what signals our actions send the algorithm. Most of all, I wish Clegg didn't blame users for the problems his company is known to cause—and to profit from.

For the foreseeable future, we're stuck in a world of greedy companies manipulating us with addictive and sadly indispensable platforms that are rife with internet rot. Let's at least learn how to stand up for ourselves and enhance our online experiences to the extent we can. We've gotten some glimpses of how to do this in this section, and we'll go deeper in the following sections.

Back to TikTok

TikTok uses an algorithm to determine which of the billions of videos on the platform to show to each of its one billion users. Thanks to its ability to surface videos that appeal to each user's individual tastes, often with uncannily personal insight, this algorithm is sometimes said to read your mind. How does it do this?

The New York Times hunted for answers and got hold of an

internal document labeled "TikTok Algo 101." Written by a TikTok engineering team, this document was used internally to explain the basics of the algorithm to other employees. In a December 2021 article, the *Times* wrote that this for-TikTok's-eyes-only document includes a "rough equation for how videos are scored . . . : *Plike* × *Vlike* + *Pcomment* × *Vcomment* + *Eplaytime* × *Vplaytime* + *Pplay* × *Vplay*." The article made no attempt to explain this awkwardly typeset formula, let alone any of the symbols in it, beyond stating that "a prediction driven by machine learning and actual user behavior are summed up for each of three bits of data: likes, comments and playtime, as well as an indication that the video has been played." Frankly, this explication is mathematically (and grammatically) questionable gobbledygook. But having already seen Facebook's formula, I'm pretty confident that I can decipher TikTok's. Perhaps at this point you can, too.

Surely, *Plike* in TikTok's formula is the estimated probability that the user clicks the heart-shaped Like button on the video, while *Vlike* is the point value engineers have assigned to this form of engagement. Same story for *Pcomment*, commenting on a video, and *Pplay*, playing a video. *Eplaytime* is a bit different, since play time is not a discrete event that either happens or not, like clicking the Like button or leaving a comment. Instead of the algorithm predicting a probability for this type of engagement, it predicts the expected number of seconds the user will watch the video for. That, I have no doubt, is what *Eplaytime* indicates, and *Vplaytime* must be the point score indicating how many points each second of play time is worth. If, hypothetically, a comment is worth twenty points and a second of play time is worth two points, then a 50 percent chance of commenting would count for the same amount of engagement as an expected five seconds of play time. With this understanding, we see that the TikTok formula is a weighted sum of engagement predictions—just like the formula central to Facebook's algorithm.

The secret TikTok document goes on to explain that "the recommender system gives scores to all the videos based on this equation, and returns to users videos with the highest scores." Sound familiar? Yes, TikTok's mind-reading algorithm is, at its mathematical core, nearly identical to Facebook's algorithm. Both rank posts/videos according to a weighted sum of the amount of engagement they are expected to elicit from the user. The TikTok document clarifies that "the equation shown in this doc is highly simplified. The actual equation in use is much more complicated, but the logic behind is the same." It's this "logic" behind the formula that we can all use to up our TikTok game, despite knowing neither the full formula nor the engagement point values.

Have you ever seen a video with overlaid text saying something like "Wait for it," "You won't believe what happens," or "You've gotta watch till the end lol"? These phrases tend to bump up the expected play time for most users, so it's a cheap trick to boost the video's score. If you don't want this kind of clickbait filling your For You page, scroll on by: don't give these videos the points they're trying to manipulate you into providing.

Some people post videos where literally nothing happens,

but they trick you into watching multiple times, thereby racking up even more expected seconds. The math is simple: if the algorithm thinks you'll watch a ten-second video three times, that's thirty seconds of expected play time. This might be a video of a crowd with overlaid text like "Look how that one guy embarrassed himself." You watch over and over hoping to find that "one guy" in the crowd, even if he doesn't exist. Or it might be a riddle that nobody can solve. All the seconds people spend watching the video over and over, hoping that with one more viewing they'll crack the nut—and all the comments they leave offering potential explanations or simply expressing frustration—are points in the engagement formula. And that's why the enigmatically nonsensical nursing assistant video did so well. It tricked users into providing a huge amount of engagement, which in turn tricked the TikTok algorithm into showing the video to millions of users.

You can't stop wannabe influencers from using ploys like these, but you can be mindful about the signals you send the algorithm. If you don't like a video for whatever reason, limit your play time. Importantly, resist the urge to rewatch it out of frustration or disgust. And don't give in to the temptation to comment. Comments and seconds watched, no matter what quality and kind, tell TikTok's algorithm one thing: "Give me more videos like this."

The Platform Formerly Known as Twitter

The dominoes of algorithmic secrecy have continued to fall. Before Elon Musk purchased Twitter for $44 billion in October

2022 (and renamed it X in July 2023), he polled his followers asking whether Twitter should make its algorithm open-source. A resounding 83 percent said yes. In March 2023, a big chunk of Twitter's source code, the computer instructions behind the algorithm, was posted online. Technically, *open-source* means more than just sharing code online; it's a whole ecosystem in which anyone is allowed to suggest edits and additions to the code. But posting the code publicly was still a huge step toward openness—and an unprecedented one for the big social media platforms.

Musk prefaced the code release by saying it was going to be "quite embarrassing, and people are going to find a lot of mistakes" and "discover many silly things." He further added, "Even if you don't agree with something, at least you'll know why it's there, and that you're not being secretly manipulated." Right away, an eagle-eyed engineer at Meta noticed a curious detail that definitely fell in the embarrassing/silly category: Twitter's code identified all tweets by Elon Musk himself and elevated them in the feeds of all users. In other words, the code was rigged to automatically amplify Musk's tweets. Journalists uncovered the rather adolescent backstory to this. After a tweet by Musk during the 2023 Super Bowl failed to generate as much engagement as one by President Biden, a furious Musk called the Twitter engineering team in for an emergency coding session to doctor the algorithm in his favor.

While the code in all its gory details is quite complex, at its core is a structure you'll find very familiar by now: a weighted sum of estimated engagement probabilities used for the bulk of the ranking process. For all the jousting by the tech giants, for all the competition to build the best social media platform, it turns out that Facebook, TikTok, and Twitter all run on essentially the same simple math formula. With Twitter, finally, we have the formula in full view as all the forms of engagement *and* the weights are known, thanks to the code release.

In Twitter's formula, a Like is worth one point. A retweet is two points. Replying to a tweet is fifty-four points. Replying to a tweet and getting engagement on your reply from the author of the original tweet is 150 points. Clicking into the conversation of a tweet and staying there for at least two minutes is twenty-two points. Clicking into the conversation of a tweet and Liking or replying to a tweet there is twenty-two points. Opening the profile of the tweet's author and Liking or replying to a tweet there is twenty-four points. Watching at least half of a video in a tweet is worth one hundredth of a point. (With such a minuscule weight, I don't know why they even bother with that one.) There are also engagements with negative weights. Clicking "Not interested in this post" or blocking/muting the author of a tweet is –148 points. Reporting a tweet is a whopping –738 points. I told you reporting posts about the Kardashians would keep you from seeing posts about them.

A few observations about these weights. First, retweets aren't worth as much as you might expect. An account with a million followers once retweeted one of my posts, and I was surprised how little impact this had on my tweet's views, but now I see why. Second, the negative weights are quite high. I never bothered clicking "Not interested in this post" before, but now I'll be sure to use that option. Third, looking at a post rather than scrolling past it is not counted in Twitter's engagement formula the way it is in Facebook's. On Twitter you can freely rubberneck at whatever posts you want without sending the algorithm the wrong signal.

There's no easy way to extract a recipe for virality from Twitter's disclosure of engagement types and their corresponding weights. But the main takeaway is that if you want your tweet to get lots of views, try to get people to write replies to your tweet and to other tweets in the conversation your tweet is part of. That's the form of engagement that matters the most—much more than Likes and retweets.

Be warned that engagement weights aren't the whole story for Twitter. For instance, paid subscribers get boosts to their posts, though presumably not as big as the boost Musk gave himself. And, like with the other platforms, plenty of other adjustments to the rankings are made after the initial weighted sum is calculated. We also don't know if Twitter has changed the weights since releasing this version of the code in March 2023. And Musk didn't release any of the code behind the algorithm that places ads in users' feeds. That said, it's fascinating to see what is emphasized with Twitter's engagement weights and to finally get a fuller view of the math behind a popular social media platform. What strikes me most is how arbitrary these point systems are. Yet nearly all of us play the social media game to some extent, so we might as well learn the rules as best we can.

You Are What You Eat

The weighted-sum formulas behind social media tell us that Nick Clegg was right: it takes two to tango. What we see online is dictated both by the platform mechanics that are out of our hands and by the trails of engagement crumbs that we leave for the algorithms. I wish that the enormously profitable tech titans would work harder to make our online experiences more positive, pleasant, and productive. But, as I'll discuss in more detail in the next chapter, their profit motive drives them to keep us online as long as possible, no matter how toxic or mindless that time is. Knowing that it's not in the financial interests of these companies to help us out, and absent laws forcing them to do so, for now it's on us users to improve our own social media experiences. The overarching lesson from our discussion of the secret formula behind social media is that we can develop healthier feeds, but doing so takes restraint and intentionality.

Imagine there's a KFC in your town, and one time after a stressful day at work you give in to temptation and head there for an easy dinner. The next day, the KFC has mysteriously moved one block closer to your house. Now the convenience and the allure are even greater, so you find yourself visiting it more often. But each time you do, the KFC moves even closer to your house. Soon it's down the block from you and is part of your weekly routine. Eventually the KFC is next door, and you're eating fried chicken more often than any reasonable human being should. You're not proud of it, but how can you resist when KFC is the first thing you see (and smell) in the morning and the last thing before bed at night? This is how social media algorithms work. They bring the things we engage with closer and closer. Once we start clicking the social media equivalent of junk food, we're going to be served up a lot more of it—which makes it harder to resist. So we click it more and the algorithm promotes it even more highly in our feeds. It's a vicious cycle that can quickly turn our feeds into endless streams of digital dreck.

My wife used to feel a lot of anxiety when spending time on social media. She'd try to get the scoop on breaking news stories on Twitter before they hit the mainstream press. But there was so much vitriolic arguing, superficial dunking, and brazen prevaricating that the whole thing became an exercise in aggravation. Eventually she quit the app and switched to Instagram, in part to help promote a flower farm business she started in our backyard. I was writing this book chapter at the time and couldn't resist discussing social media engagement formulas with her. Curious, she tried training her feed to be nothing but flowers and fluffy dogs, and it worked. Now when she goes online, she feels a sense of calm and comfort instead of frazzled rage. You, too, can shape your social media experience in whatever ways you want, now that you know the secret formula.

8

Standing Up to
the Tech Titans

Throughout this book I've tried to focus on math that you can use to navigate the challenges and decisions you face in life. In this final chapter I'll continue to provide practical hands-on math lessons, but I'll also use math as a tool to tell an important story about the rise of the tech giants and what could be done to address some of the imbalances that have arisen in our economy. While at times this entails leaning more heavily into advocacy than I've allowed myself in the preceding chapters, I think it's important to see how math can guide not just our individual actions but our collective ones, too.

As of August 2024, the four richest people in the world are tech moguls: Elon Musk, Jeff Bezos, Mark Zuckerberg, and Larry Ellison. Six of the seven largest companies by market capitalization—which, you may recall from chapter 2, is the total value of a company's shares traded on the stock market—are tech companies: Microsoft, Apple, Alphabet (the parent company of Google and YouTube), Amazon, Nvidia (the chip maker powering much of today's AI), and Meta (the parent company of Facebook, Instagram, and WhatsApp). There's no denying it: tech titans rule the economy. And there's nothing inherently wrong with this. Tech is all about innovation and giving us tools and toys that just a few years ago didn't seem even remotely possible. But look deeper and some serious problems emerge in this saga of the late twentieth- and early twenty-first-century rise of big tech.

The tech giants have become so large and powerful, and so central to our lives, that we keep using their products even when they degrade to crappy, ad-infested shadows of their former selves. Google's bottom-line business is getting us to view and click on ads, and it shows. Google search has filled up with far more sponsored content than it ever had before, making it harder to find what we're looking for. Amazon operated at a loss for many years, fueled by zero-percent interest rates and venture capitalists who prioritize growth and "disruption" above all else. This put competitors in a difficult bind: join Amazon's marketplace, or resist and get crushed by the online behemoth. Now comfortably perched atop the e-commerce throne, Amazon has been filling its site with sponsored content and sneakily driving customers to its corporate partners, whose products are often overpriced and of lower quality. When you're the only game in town, you set the rules with impunity and make sure you're the winner—even when doing so means your customers are the losers.

If your favorite brand of milk or eggs becomes too expensive or stops tasting good, you switch brands. When was the last time you considered switching from searching the web with Google or shopping on Amazon? If your answer is never, don't feel bad—it's not your fault. Alternatives are hard to find. That's because the tech giants frequently acquire competitors to preserve their market dominance, and because startups have little chance of competing with the vast computational resources and data troves the incumbents have exclusive access to. I use Google and Amazon all the time, even though I'm often frustrated with them. But I have an idea for clawing back some of the power the tech industry wields over us.

I believe we can use math, some of the same math the tech titans have been riding to the top, to push back and resist the ways these companies use their algorithms to play us like pawns and manipulate us for profit.

Lost in the Amazon Jungle

Julia Angwin is a Pulitzer Prize–winning journalist focusing on the societal impact of technology, currently as an opinion writer for *The New York Times*. Angwin was a math major in college, and she didn't abandon her youthful interest in mathematics when she turned to journalism. Quite the opposite: she carried this interest with her and wed the two disciplines in a creative and productive way. In 2018, Angwin cofounded a nonprofit news organization called *The Markup* that specializes in data-driven journalism. This means investigations based less on hunches and interviews and more on cold, hard numbers. Several of the anecdotes of numerification and invasive data surveillance sprinkled across this book's Kafkaesque opening scene were drawn from *Markup* investigations. The one I want to tell you about here concerns Amazon and its opaque algorithm.

Shopping online is very different from shopping in a brick-and-mortar store where all items for sale are readily visible on the shelves. On Amazon, the algorithm that ranks search results shapes what you see—and hence what you buy. One ex-employee said that 70 percent of customers never click past the first page of results, and 35 percent of customers click the very first search result. In 2019, Amazon told Congress that the company's algorithm does not consider whether products are sold by Amazon-owned brands. But an investigation by *The Markup* in 2021, when Angwin was serving as editor in chief, seems to suggest otherwise.

The Markup analyzed nearly 3,500 popular search queries on Amazon. It found that in 60 percent of them, the item listed in the coveted number-one spot wasn't an organic search result; it was an ad—meaning a product the seller paid to be shown prominently. These ads were listed as sponsored, so at least if the customer was paying attention, they could recognize these for what they were: spots sold to the highest bidder, rather than earned through ratings, reviews, and sales.

More shocking is what *The Markup* found in searches where the top spot was *not* listed as sponsored. Half of these gave the top spot to an Amazon-owned brand or a brand that sells exclusively on Amazon—even though such brands counted for only

6 percent of the products seen in these searches. Among the searches where a product sold by an Amazon brand or exclusive secured the top spot, such products were placed above better-rated non-Amazon products with more reviews over a quarter of the time. (Amazon doesn't publicly list sales numbers, but a product's number of reviews serves as a rough proxy.)

To further probe the situation, *The Markup*'s investigative team trained a machine learning algorithm to predict which item would land in the number-one spot in searches. Of the half-dozen factors they tested, including star rating and number of reviews, the factor with the biggest impact was whether the item was an Amazon product, meaning sold by an Amazon-owned brand or a brand that sells exclusively on Amazon. I love the delicious irony here that Amazon, like so many other tech-oriented companies nowadays, uses sophisticated math in the form of machine learning to predict and influence customer behavior—and *The Markup* used some of the very same math to catch Amazon in this unscrupulous act of market manipulation.

Angwin told me that this Amazon investigation "pushed the boundaries with math" because of her team's use of machine learning. "Our job as journalists," she explained, "is to look at the world and tell you what's happening. So I'm always a little wary of the realm of prediction. That's the realm of pundits. But in this particular case, I realized that the only way we could make sense of this data was to use this probabilistic approach." Angwin was referring to *The Markup*'s machine learning algorithms that predict which items will top various search rankings and in doing so reveal which factors the Amazon algorithm seems to prioritize. She described this approach as "a bit of a risk" but said the response to the article was really strong. Congress wrote a letter to Amazon asking the company to explain *The Markup*'s findings, since they seem to contradict the company's earlier testimony. Amazon's response was not made public, but Angwin said the Congressional committee "decided the response was not adequate" and referred the matter to the Department of Justice for a perjury investigation. "I felt so vindicated that this math that was a little risky was actually really convincing," she told me.

To see what all this looks like in practice, consider breakfast cereal. You might expect to find familiar items like Cap'n Crunch and Honey Nut Cheerios atop Amazon's search results. These both have five-star ratings and over ten thousand customer reviews. But *The Markup* found them placed below a cereal sold by the Amazon-owned brand Happy Belly, despite it having a comparatively meager four-star rating and one-tenth as many reviews.

Some of these Amazon-owned brands have been accused of cloning popular products. Amazon settled a lawsuit with Williams-Sonoma that claimed, among other things, that Amazon copied West Elm furniture and sold knockoff versions under an Amazon brand. The CEO of Allbirds, a shoe and clothing company with a focus on sustainability, wrote an open letter accusing Amazon of copying one of his company's shoe designs and said that if Amazon is going to do this, it should at least copy Allbirds' sustainability practices as well. To eat into the market share of Amazon's competitors, these knockoffs don't have to be cheaper than the originals. They just need to show up higher in the search rankings, since that's what drives sales on the platform. How could Amazon resist pulling the levers of its algorithm to boost its own brands?

The problem is that Amazon competes with other retailers while running the marketplace in which this competition takes place. It would be like having all the referees in the NFL employed by the New England Patriots rather than being neutral and independent. In retrospect, we were all foolish to think this could have turned out well. And, dismayingly, this hurts both customers and retailers.

Customers are drawn to inferior products, purchasing lower-quality items that cost more because the invisible hand of the Amazon algorithm guides them to do so. Fortunately, once you are aware of this, it is not hard to resist the algorithmic influence and beat Amazon at its own numbers game. Simply click the drop-down menu to switch the search rankings from the default setting of "Featured"—a conspicuously vague term that in essence means letting the algorithm sort however Amazon wishes it to do so—to one of the other options, like average customer review or lowest to highest price. One study found that users who choose an item from the top four Featured spots end up paying 25 percent more than the best deal. On average, you need to scroll down to the seventeenth item in the default Featured rankings to find the best deal. Even with the other options for sorting search results, don't trust the algorithm: check prices and reviews to make sure you're not getting duped.

Retailers, on the other hand, face an unenviable dilemma.

They can compete against Amazon outside the platform, where the odds are stacked against them. Or they can compete against Amazon within the platform, where the algorithm is stacked against them. It's no wonder that many sellers resort to spending large sums to have their products placed higher as sponsored listings. But even this is no guarantee that a given product will stand out among the crowd. A study in 2021 found that the first page of results on Amazon searches contained an average of nine sponsored listings, more than twice the number Walmart had at the time.

It wasn't always like this. A decade ago, Amazon's search engine was an efficient machine for surfacing good deals. Now, sponsored products are squeezed into every nook and cranny of the website, often in insidious ways that make it hard to tell what's real and what's just an ad. Finding organic content on Amazon is becoming as hard as finding organic produce at 7-Eleven.

Unfortunately, I don't have any suggestions for how retailers can fight back against the unfair Amazon algorithm the way customers can by re-sorting the rankings and carefully comparison shopping. But the sci-fi author Cory Doctorow has a clever idea for how we could reclaim some of Amazon's market dominance. While his idea involves a little imagination and a little math, it isn't science fiction. Allow me to explain.

The convenience of Amazon's online shopping experience is undeniable. But so is the harm it causes to the brick-and-mortar shops that many of us love. It would be nice to have the power of Amazon's website to find products and reviews, and its gargantuan inventory, without taking business away from local shops. For books, there is a slick web browser plug-in that helps. You can shop for books on Amazon as usual, and while doing so the plug-in automatically shows which libraries near you have the book you're looking at. And it lets you easily click to reserve books from these libraries instead of buying from Amazon. Doctorow notes that it would be simple to expand this plug-in to include an option for purchasing from local bookstores. And, as he writes, why stop there?

If the ISBN numbering system for books were extended to all items, it would be straightforward to expand this browser plug-in even further to cover all online shopping. This would put a much bigger dent in Amazon's monopolistic powers. Nobody has done this yet, in part because building a numbering system flexible enough to cover all conceivable items, including those not yet invented, is challenging. But there's an entire field of mathematics devoted to enumerating objects, called combinatorics, and I have no doubt the numbering system we need for the browser plug-in Doctorow envisions is possible. This plug-in would help customers shop locally and support independent businesses by cutting out Amazon's exorbitant commissions, which have been estimated to reach up to 50 percent. In short, it would co-opt Amazon's Goliath site and give power back to the Davids of retail commerce.

All of this provides the perfect illustration of what I mean by Robin Hood math—from the very simple idea of sorting search results by ratings or reviews, to the more ambitious idea of extending ISBNs to all products, to the more sophisticated idea of using machine learning and statistical analyses the way *The Markup* did to uncover signs of unfair market manipulation. Amazon may use math to rig the numbers game behind its influential algorithm at our expense, but we can use math to fight back and empower ourselves.

There are no laws forcing retailers like Amazon to limit the fraction of listings on each page of search results that are sponsored. And American laws don't adequately address the unfair advantage a company like Amazon gets from controlling the dominant marketplace on which its own products compete with those of other retailers. Until we get more help from lawmakers, we should use all the help we can get from math.

Searching for Answers

Google is practically synonymous with web search. It's tempting, then, to think of Google as a web search company, one that happens to also sell phones and provide email and other online services. A more accurate description, however, is that Google is an advertising company that happens to provide a web search service among other things. The reason I say this is because nearly 80 percent of Google's $300 billion in annual revenue comes from advertising.

Online advertising accounts for the majority of all global advertising, nearly tripling the next-largest medium, TV. And Google is the largest online advertiser in the world. Next on the list is Meta, then the Chinese tech conglomerate Alibaba. Coming in fourth, but with one of the fastest growing advertising revenues, is Amazon. That's thanks largely to the proliferation of sponsored products that we just discussed; those now account for nearly 80 percent of Amazon's $40 billion in annual ad revenue.

The majority of Google's advertising money comes from ads placed on web searches. It's wonderful that we all get to use this powerful tool for free, but let's not be naive about what happened. In its earlier days, Google had to establish its web search as a superior product to secure market dominance. But once it did so and became comfortably ensconced as the search engine of choice for billions of people, Google no longer had to fear users leaving for alternatives and so could ramp up revenue by doing the same thing Amazon did: filling its pages of search results with ads. And it's not just the official ads labeled Sponsored that are the problem. The livelihood of many businesses depends on their ability to land on the coveted first page of search results, so there are endless attempts to game Google's rankings. The net result is that it has become harder to find quality information on Google.

In 2020, Geoffrey Fowler, a *Washington Post* technology columnist, showed that this degradation of Google isn't just in our imagination. He searched for T-shirts and found that the first organic search result was relegated to the ninth row of content on Google's page. In addition to ads, Google buried the search results under things like "snippets" that attempt to extract information from other sites, and links to Google-owned sites such as Maps and YouTube. Those snippets and links are framed as a convenience, but they're really tricks for keeping you, and the ad revenue you generate, within Google's walled garden. Fowler writes that "on some searches, it's like Where's Waldo but for information. Without us even realizing it, the Internet's most-used website has been getting worse." To document the change, he used the Internet Archive's Wayback Machine, which saves older versions of websites, to see what search on Google looked like decades ago. He found that in 2020 you needed to scroll down six times farther than you did in 2000 to find the first unpaid link to an outside website. One study of ten thousand different searches found that in 2013 organic search results started an average of 375 pixels down the site, whereas by 2020 that number had nearly doubled to 616 pixels.

There's no simple drop-down menu to improve the quality of search rankings on Google like there is on Amazon, because information is a lot harder to rank than products that have straightforward metrics like customer ratings. But clicking the Web tab at the top of a Google search gets rid of the AI summaries, sponsored links, and other clutter that now crowd the top of the results page. Additionally, there is a semi-mathematical syntax for search queries that can help narrow down search results. Most people don't know about this, so let's take a look.

You can use a minus sign (the hyphen key) to keep words or phrases out of your search results. Looking for square-shaped pants but your results are flooded with a certain undersea cartoon? Try searching **square pants -SpongeBob**. You can use two dots (periods) to search for numbers within a range. Looking to read about a war that occurred in Europe in the early

1800s but you forgot what it was called and when it took place? Searching **Europe war 1800..1820** returns results that includes any number between 1800 and 1820. You can also tell Google to fill in the blanks with the asterisk operator: searching **famous quote "the best * is *"** will help you find expressions like "The best defense is a good offense" and "The best fighter is never angry."

Government websites can be notoriously difficult to navigate. Many include bare-bones search functions, while others have no search features at all. Not to worry. You can tell Google to search a specific site by appending the site operator: searching **COVID vaccine site:cdc.gov** will return pages on the CDC's site that contain information about COVID vaccines. If you're looking for a specific file format, like a PDF, add **filetype:pdf** to your search. Remember seeing a cool PowerPoint presentation about the expected value formula but having trouble finding it? Try searching **expected value formula filetype:pptx**.

Some people have noticed that AI-generated images are taking over image search results, which is infuriating when you're looking for real stuff instead of synthetic fakery. Appending **-AI** and **-prompt** to your search isn't perfect but it tends to help eliminate a lot of these. For a more extreme measure, you can restrict the date to before AI image generators took off—for instance, appending **before:2022-4-1** should work, though of course this isn't helpful if what you're looking for is more recent.

Beyond this, I do have an idea for how math could help reduce the amount of garbage we see online. But it requires a change

Google would have to make internally, not something us users can do from the outside. I proposed it in an op-ed in *Slate*, but so far Google hasn't expressed any interest. That might have to do with the fact that Google makes money off this garbage and has no real incentive to clean up the mess. To understand this story, we must look beyond web search to a less visible way that Google makes money from ads.

When you visit a site like cnn.com or foxnews.com, you see various ads in various locations. In some very rare instances the owner of the site directly negotiates with advertisers the rates and placements of these ads. But far more often this is done through an intermediary ad exchange. Think of it this way. If you have a website that's drawing some traffic that you'd like to monetize, you do so by renting out space on your site to an advertiser, the same way that someone with property next to a busy highway might rent space on a billboard. But it would be difficult and time-consuming to find interested advertisers yourself. Meanwhile, advertisers want to place their ads on sites across the web. But they usually don't have the time and resources to find and vet the host sites and to negotiate all the contractual details. Ad exchanges match host sites with advertisers, and they do this in a very efficient, algorithmically automated way that results in millions of ads placed across millions of sites in a matter of milliseconds. Google runs the largest ad exchange in the world. Google's ad exchange is so massive and dominant that in September 2024 it became the subject of a federal antitrust case in the US.

Let's take a moment to unpack this. When you search Google for bicycles, it's not surprising that you see ads for bicycles and that Google makes money from showing you these ads. But when you visit any website on the internet that has ads, there's a good chance Google is the company placing those ads. And when it does, it takes a slice of the ad revenue, similar to how realtors charge a commission for connecting home buyers with home sellers.

The problem is that people have figured out that one way to generate traffic, and hence ad revenue, is by publishing fake news. Most advertisers don't want their ads placed on harmful sites, but since they purchase ad placements through an automated exchange, they usually don't know where their ads end up unless they go to significant effort to find out. This has led to shocking situations like the World Health Organization advertising on anti-vax sites. Not only is this poor optics, but it means the WHO is unintentionally sending money to organizations that undercut its public health efforts.

Google has policies prohibiting certain categories of misinformation, both in ads and in the host sites that display the ads. But studies have found a huge number of fake-news sites receiving ad placements nonetheless. One report estimated that in 2021, fake-news sites earned over $2.5 billion in ad revenue. Not all of this came from ads placed by Google, but other investigations found that a sizable fraction of ads on fake-news sites did come by way of Google's ad exchange. When I looked into this, I found something fishy that I suspect is part of the problem.

Google's ad placement policies say that when a specific URL is found to contain any of the prohibited types of misinformation, it loses the ability to host ads from Google. When a bunch of URLs belonging to a single website are demonetized in this way, the entire website can lose the ability to host ads. At first blush, this sounds reasonable. But what's strange is that this all-or-nothing approach is not how content moderation works in many other settings—because it's known to be ineffective and problematic.

Social media sites like Facebook and Instagram do sometimes delete posts that violate their policies, but more often they demote posts instead. Here's how that works. There are far too many posts to inspect manually, so AI systems do the bulk of content moderation. These systems aren't reliable enough to decide with certainty whether a given post violates any policies, so they produce an estimate of the likelihood of a policy violation. This estimate is factored into the ranking formula we learned about in chapter 7, so that the more likely the AI judges a post to violate a policy, the lower in everyone's feeds the post appears. This reduces the amount of bad stuff people encounter on social media without having to make hard black-and-white decisions of which posts stay up and which posts are taken down.

Google uses this flexible approach to content moderation in its search ranking algorithm. Indeed, Google is very reluctant to delist webpages entirely, but it also doesn't want people seeing harmful search results, so it uses AI scoring systems to push bad stuff lower in the pages of results. In my *Slate* article, I argue that Google should add a penalty factor to the algorithmic auctions it runs in its ad exchange, similar to the penalty factor it uses in search and that most social media platforms use. The more confident Google's AI system is that a site contains prohibited misinformation, or any other policy violations, the more costly—and the less likely—it should be for the site to land an ad placement, even if the AI isn't confident enough to demonetize the site.

Why doesn't Google already do this? I suspect the answer is that Google simply has no incentive to improve its ad-policy enforcement. In fact, it is financially incentivized to enforce its ad policies in a minimal, feckless manner. Here's why. When you see something bad on TikTok, you get mad at TikTok. When you see something bad on Google, you get mad at Google. Public scrutiny motivates these sites to clean up their acts, at least to an extent. But when you encounter a dangerous fake-news site, you get mad at the site without realizing that another company, quite possibly Google, is funding the site with ad revenue. Google suffers no reputational damage when it fails to enforce its ad-distribution policies, and it profits from all ad placements, even the harmful ones. So why should we expect it to do better?

Frustratingly, the punchline seems to be that we know how to use math to more effectively defund a lot of bad stuff on the internet, but the ad distribution companies like Google aren't going to bother until we force them to do so. And while billions of dollars funneling into the fake-news industry every year is a big problem, it's not a problem most people are aware of—so it's not a priority for lawmakers to address. Perhaps by reading this chapter and writing to your representative, you can help change that.

The Ills of Online Ads

There are essentially just two ways to pay for services on the internet: subscriptions and ads. The internet has largely gone with the ad-based funding model. Other than a few exceptions like paywalled news sites and streaming apps, most of what we encounter online is provided for free and funded through ads. Blogs and websites are free, social media and video platforms are free, Google's services are free. This sounds like a good thing, and in many ways it is. However, some experts have referred to this reliance on ads as "the internet's original sin." Since it's not so obvious how offering things for free could be bad, I'd like to explain a few of the overarching problems with ad funding—and explore a proposal to nudge the internet in a healthier direction. While the problems and the proposal are not inherently mathematical, I'll bring math into the conversation to help clarify what's going on.

First, online ads aren't very effective or lucrative. For a platform to generate substantial profits through ad revenue, scale is essential: the platform must show lots of ads to lots of users. Practically all businesses want to grow, but with online ad companies like Google and Meta, there has been a relentless drive to grow as large as possible, as quickly as possible. It's not hard to do the math here. Instagram, for instance, earned about $60 billion in ad revenue in 2023 from roughly two billion active users. These users spent an average of thirty minutes per day on the platform. That means, on average, a full hour scrolling through content on Instagram generates Meta a mere sixteen cents of ad revenue. The growth-at-all-costs business model needed for these paltry sums to add up to big profits has had some disastrous consequences.

To see one striking example, consider Facebook in 2013. The company hatched a plan to grow its user base by subsidizing mobile internet access in a handful of countries where internet infrastructure was lacking and cellular data rates were expensive. The catch: Facebook's free internet access was largely limited to Facebook itself. One of the countries targeted particularly heavily with this plan was the Philippines. And it worked. Facebook became the de facto internet in the Philippines. Even a decade later, nearly 95 percent of all internet users in the country are on Facebook.

While this rapid growth was successful for Facebook, it carried a tremendous price for society. Rodrigo Duterte successfully ran for president two years after the free Facebook initiative, skillfully leveraging the platform by flooding a freshly wired electorate with memes and misinformation. Many people couldn't afford internet access outside of Facebook's free ad-funded internet, so their primary window into the world was Facebook— where engagement, not accuracy, drives the algorithm, as we saw in chapter 7. Once in office, Duterte's brutal regime killed thousands of people in a massacre masquerading as a war on drugs, among other atrocities. One of the most poignant interviews I've ever heard is the tech journalist Kara Swisher interviewing Nobel Peace Prize winner Maria Ressa. They discuss

Ressa's courageous work running a news outlet in the Philippines that stands up to Duterte, and the damning role Facebook has played in her country.

Facebook has three billion users. Instagram has two billion. TikTok has over a billion. Google handles almost ten billion search queries every day. Big tech is built on big scale, as the math of ad revenue necessitates. There has recently been interest in more small-scale, neighborhood-like spaces on the internet. The idea is that these might be more civil and pleasant than placing billions of strangers in a giant room where whoever gets the most engagement gets the biggest megaphone. But it's hard to make this work with the economics of the internet, which for better and worse has largely been tantamount to the economics of ads.

So that's the first problem: ads drive a growth-at-all-costs mindset at the tech platforms, which has had some very destabilizing impacts on society. And once platforms reach a gargantuan scale with a devotedly locked-in user base, they often let their services degrade knowing that few alternatives exist and few users will switch to the ones that do. This degradation is partly due to complacency: the competitive drive to innovate dissipates when there is no real competition. But it's also due to another unfortunate consequence of the ad-based funding model: If users aren't going to leave anyway, why not squeeze out more revenue by cramming in more and more ads? We discussed this above with Amazon and Google, but frustratingly it's a widespread phenomenon in the tech industry. It's essentially what Cory Doctorow, the sci-fi author we encountered earlier,

termed "enshittification" in a recent viral essay. His essay reso-
nated with so many people that the American Dialect Society
chose *enshittification* as its Word of the Year in 2023.

This brings us to the second major problem with online ads.
Since they bear so little fruit on an individual basis, there's an
enormous incentive to target them as effectively as possible.
This sounds perfectly reasonable—why not show people ads
they're more inclined to click?—until you realize that this is the
main reason for the invasive tracking, surveillance, and data
collection we touched upon in the opening pages of the book.
Throughout the past two decades, the tech giants have figured
out that the way to increase the odds of someone clicking an ad
is to learn as many intimate details about the person as possible so that the right ad can be selected for them in the ad exchange. Don't like your personal data being collected when you
drive your car, study in college, search and shop online, browse
the web, or spend time on social media? You largely have the
internet's reliance on targeted advertising to blame. That, and an
embarrassing lack of data privacy laws in some countries, nota-
bly the US.

Now that a massive digital infrastructure and economy has
developed around data surveillance, one of the scarier developments potentially on the horizon is personalized pricing. Customer data—the kind that's slurped up when you shop online
and in stores with customer loyalty cards—could be used to estimate the maximum price each person is willing to pay for each
product. Imagine having to pay more than your next-door neigh-
bor, or even your spouse, for a dinner delivery just because your

delivery app knows your detailed spending history and uses it against you. And imagine Amazon buying your Google search history, a log of your Instagram activities, and all the numbers in your grocery store's loyalty card—so that Amazon can price gouge you with unprecedented algorithmic efficiency. This may sound like a dystopian sci-fi scenario, but it's not as far-fetched as you might think. Lina Khan, the chair of the Federal Trade Commission, said that "through the enormous amount of behavioral and individualized data that these data brokers and other firms have been collecting, we're now in an environment that technologically it actually is much more possible to be serving every individual person an individual price based on everything they know about you."

A third problem with ad-based funding is that it encourages platforms to maximize the amount of time users spend on a platform, and the amount of content they see when they do. That's because doing so maximizes the number of ads seen, and hence the amount of ad revenue generated. But it's often unpleasant for individuals and unhealthy for society.

For subscription streaming services like Max and Paramount Plus, and for subscription news sites like *The New York Times* and *The Wall Street Journal*, the goal is to keep customers sufficiently satisfied that they keep paying the monthly subscription cost. It doesn't matter if people log on for hours every day or only minutes each month. If people feel the product is worth the price and keep paying, then both company and customer are happy. With ad-funded free services like YouTube, TikTok, Facebook, and Instagram, profit is driven by quantity, not quality. Have you ever found yourself mindlessly scrolling through endless content on social media without enjoying it but without being able to stop? It's not your fault; it's the economics of ads. As long as social media is funded by ads instead of subscriptions, it will be engineered to be as addictive as possible and to flood you with as much content as possible.

There is much to love about having so many free online services. And charging for access to information and social connection would create some terrible societal inequities. I don't think we should ditch ads entirely and bring subscription fees into everything we do online. But I do think we need to be realistic about the downsides of ads and the harmful incentives they come with.

In economics, *externalities* are the costs that fall on those outside an organization providing a good or service. Internal costs are typically things like materials, equipment, and labor, while externalities can be anything from the pollution caused by industrial production to the suffering imposed on animals in factory farming. The harms associated with online advertising can be considered externalities. These include the pressure for platforms to host more users and content than they can safely monitor, the invasion of privacy from the excessive surveillance and tracking needed for microtargeting, and the unhealthy addiction individuals face from a focus on quantity rather than quality when it comes to time spent online.

Economists have a tool for internalizing external costs, which means making companies pay for their externalities: a Pigouvian tax. Perhaps the most familiar example of this mechanism is a carbon tax, which puts a price on carbon emissions. What's particularly appealing about a Pigouvian tax is that it encourages companies to find ways of reducing the harm the tax targets while also generating revenue that can be put toward addressing the harm. A carbon tax creates a financial incentive for companies to lower their carbon emissions, since they'll save money by doing so, and at the same time the government can spend revenue from the tax on things like carbon-capture technology. To me, this is the right way to address the harms of online advertising. I don't think we'd be better off if we outlawed online ads and forced all internet services to charge monthly fees. But I also don't think we should continue letting the tech titans earn hundreds of billions of dollars every year from online ads without paying a penny in return for the harm they cause.

One of the first people to suggest a Pigouvian tax to address the internet's original sin—a sin tax for the internet, if you will—was Paul Romer, the Nobel Prize–winning economist whose rule-of-thumb recipe for pricing risk we studied in chapter 5. He proposed the idea in a 2019 *New York Times* op-ed, then elaborated on it in a 2021 blog post after Maryland passed a variant of it. Romer's proposal is to levy a progressive tax on revenue from online targeted ads. *Progressive* here doesn't imply a liberal political orientation; it just means the larger the company, the higher the tax rate.

This would help with the monopolistic concentration of market power we touched on earlier in this chapter. The tax burden would fall most heavily on the established giants, like Google and Meta, making more room for new competitors to enter the market. And ad-funded tech companies would be discouraged from mergers and acquisitions since the progressive nature of the tax means the rate goes up when companies are combined. Google and YouTube would face lower ad tax rates if they were separate entities; same for Facebook and Instagram.

If the progressive ad tax rates are high enough, tech companies—at least the very large ones—might be incentivized to abandon targeted ads entirely. Romer sees this as a good thing. He wrote that when a business switches from an ad-based revenue model to an ad-free subscription model, "the success of the business would not hinge on tracking customers with ever more sophisticated surveillance techniques. A company could succeed the old-fashioned way: by delivering a service that is worth more than it costs."

In the US, there have recently been attempts to ban targeted advertising outright, to break up the tech giants through antitrust actions, and to prohibit some of the industry's most egregious tactics. I'm not against these, but I think there's much wisdom in using financial incentivization that gets at the heart of the problem rather than playing whack-a-mole with the various harms of digital advertising. And we could use the revenue from a tax like Romer's to address some of these harms.

In an article published by the Poynter Institute, the journalist Lucia Walinchus argues that an online ad tax should be used to help fund nonprofit news organizations. She says that journalism is a public good and that "a portion of all cable bills goes to C-SPAN. A portion of digital ads should go to journalists." If you believe that Google and the big social media platforms pollute our information ecosystem with misinformation and other problematic content, then it makes sense to have them contribute to cleaning up the mess. Funding public journalism strikes me as a logical component of this.

The media scholar Ethan Zuckerman (who also happens to be the inventor of the pop-up ad) has long advocated for publicly funded alternatives to commercial social media platforms. I think this is a great idea, too, and an ad tax could help fund it. I recently served on a panel discussing democracy and disinformation at the Organisation for Economic Co-operation and Development (OECD), and at one point I drew a parallel with the BBC. In the 1920s, the UK launched the BBC to provide public-service radio. As the dominant medium switched from radio to TV, the BBC followed suit by providing public-service TV stations. But now much of the world gets its information from social media, yet there is no real public-service social media platform to counterbalance the financial incentives of the for-profit corporate ones. An ad tax would help adjust these incentives and could help pay for a public-service platform, bringing institutions like the BBC into the twenty-first century.

As AI poses a growing risk of widespread job loss due to automation, we may need to take seriously the idea of a universal basic income (UBI). Perhaps an ad tax could help fund this. Much of AI's current progress has its roots in the research labs of Google and Meta, the two largest online advertising companies. Their research was motivated in no small part by the need to predict which ads users are most likely to click—a task well suited to AI. Funding a UBI with an ad tax would force Google and Meta to foot some of the bill for this labor loss, and it would discourage others like OpenAI and Microsoft from infesting their AI products with ads and going down the dark path of data surveillance.

One idea I've had for bringing a bit more mathematical precision to the ad tax proposal is to make the rate progressive not just with respect to the size of the company but also with respect to the granularity of the targeting. If you target ads based on very broad interests and demographics, you'd pay a low tax rate. The more you microtarget with very specific, personal data, the higher a rate you'd pay. I haven't worked out the details of what this might look like in practice, but it seems worth exploring further. It's a more focused way of addressing the surveillance and tracking externalities of online advertising.

Even without this extra layer of math, there's a beautiful mathematical balance to a progressive ad tax. To grasp it, we must look back to the work Paul Romer published in 1990 that earned him a Nobel Prize in 2018. It had long been understood that something was missing from the prevailing economic theories prior to 1990. Global economic growth was supposed to slow down as we depleted our finite supplies of raw materials, but data showed that growth continued and even accelerated over the centuries. Economists knew that they needed to address the role of technological innovation in propelling economic growth. But before Romer, nobody had figured out how to work that observation into the equations of macroeconomics.

One of Romer's key insights was to distinguish between inputs to production that can be used simultaneously by an unlimited number of people and those that cannot be so used. A desktop computer can process only one command at a time. Employees generally can work for only one company at a time. But a piece of software, like Microsoft Windows, can be copied an unlimited number of times and used by everyone all at once. My use of Windows has no impact on your use of it. That something can be shared widely while still being proprietary intellectual property, rather than a public good, is not specific to software. Other examples include ideas, product designs, and math formulas, all of which were previously unaccounted for in macroeconomic theory.

In his papers, Romer articulates a crucial mathematical consequence of his theory: if a company doubles all its inputs to production, and if any of these inputs are the shareable kind, then the output more than doubles. That's because doubling the nonshareable inputs alone suffices to double the output. Twice as many construction workers with access to twice the construction materials generally results in twice the rate of construction. If the construction workers also create or receive a plan that helps them operate more efficiently—an increase in the shareable inputs to their production—then their construction rate more than doubles.

This property of a quantity more than doubling when its input is doubled is an example of what's known in math as *nonlinearity*. Specifically, it is *superlinearity*, because it is growth that is faster than linear growth. (Linear functions double whenever you double the input, and the *super* here is Latin for "above.") A common example of superlinear growth is the exponential function. This is in contrast with sublinear functions, which are slower than linear. A familiar example of a sublinear function is the logarithm, as we saw in chapter 5 when exploring happiness.

The superlinearity Romer noted helps explain the incredible rise of the tech giants. One of the most valuable commodities companies like Google and Meta have is user data, which is a perfect example of a proprietary shareable resource. It is freely shared within a company and used to power all the company's algorithms and apps without any competitors having access to it. The bigger these companies grow, the more user data they collect, and the more they benefit from the superlinear value of that data. This makes it hard for other tech companies to compete with them—and hard for other industries to compete with the tech industry as a whole. Indeed, other industries have far fewer shareable inputs that lead to superlinear growth than the tech industry, where ideas, apps, algorithms, and data rule the road. In a blog post titled "What I Wish Someone Had Told Me," Sam Altman, the CEO of OpenAI, wrote that "you really want to build a business that gets a compounding advantage with scale." Romer put this kind of advantage into a macroeconomic framework, and the tech giants put it into practice.

This takes us back to the ad tax proposal. A fixed tax rate on online ad revenue would be linear: if a company grows and earns twice as much from ads, then it would pay twice as much tax on that revenue. A progressive tax, on the other hand, is superlinear: when revenue doubles, the tax burden more than doubles. The superlinear value tech companies derive from user data is better balanced by a superlinear tax on the ad revenue this user data generates. In other words, making the ad tax progressive is a way of using superlinearity to offset superlinearity—to restore some balance in our economy and reduce the tendency the tech industry has exhibited toward market concentration.

A Tool for Everyone

Leda Braga is known as the most powerful woman in hedge funds. She competed in Math Olympiads in school in her native Brazil before earning a PhD in engineering from Imperial College London. It was only at that point that she entered the world of finance, starting as a quant at JP Morgan, then working her way up to her current role as the founder and CEO of Systematica Investments, a hedge fund specializing in quantitative methods and algorithmic trading systems. Not too long ago, Braga invited me to speak at her company's annual off-site retreat. I had just written my first book, an exploration of the role data and algorithms play in the media landscape called *How Algorithms Create and Prevent Fake News*, and Braga wanted me to speak on this topic. I wasn't sure why a hedge fund would want to hear about this, and to hear a mathematician speak about it, so I asked Braga. I was fascinated by her answer.

Braga explained that the hedge fund industry has adeptly wielded math to generate wealth, helping businesses and retirement accounts grow in ways she is extremely proud of. But she feels that the more those with quantitative backgrounds benefit from math skills, the more the onus is on them to help others without these skills. She was hoping my talk would inspire her number-savvy employees to look for problems in the world that mathematics can help address, like protecting our fragile information ecosystem. I'm not sure how successful I was, but that conversation left a mark on me. It got me thinking about what else I, a math educator and writer, could do to help.

The best answer I could come up with is to write this book, showing readers of all backgrounds that math is a tool for everyone. A tool that helps us navigate the data-driven world we live in. A tool that helps us recognize and resist the ways we are manipulated by the apps and algorithms that are now woven into the fabric of our lives. A tool that has powered much of the success of those on Wall Street and in Silicon Valley, but which can benefit everyone. It's time to honor Robin Hood's spirit by taking math from the rich and giving it to the rest of us.

Epilogue

Seven a.m., your day begins. You check the TikTok video you posted yesterday and are pleased to see all the views it has garnered. After asking viewers to suggest names for the adorable doggies at the Saint Bernard rescue organization you volunteer for, you were flooded with nice comments. These comments build community for your channel. And they score points in the weighted sum that powers the algorithm, helping your video get the visibility it deserves.

For a while you've wanted to adopt one of the Saints at your rescue organization, and you've recently developed a special bond with one of the dogs there. She's such a sweetie, but she's eight years old and Saint Bernards sadly live to only nine years on average. You can't imagine bringing home a new best friend only to lose her a year later. Then you remember what you learned about life expectancies, how averages aren't the whole story. You do some online research and find many owners reporting that their Saints lived to twelve or even thirteen. You decide to look at it from a Bayesian perspective. At birth, the odds of a Saint Bernard making it to that age aren't too high. But the fact that this one is already eight and still in good health causes you to update the odds considerably, giving you more optimism in the matter.

Even so, you're on the fence as to whether it's worth the emotional investment getting deeply attached to a dog with at most a few years left. You keep the Bayesian mindset going. When you say you're on the fence, you're in essence assigning a 50 percent probability to the notion that it's a good idea to rescue an older dog. More information would help, so you talk to some friends and post on social media asking about others' experiences. (And when making this post, you include an eye-catching photo of the soulful eight-year-old Saint you're considering, knowing this will help get more engagement, hence more promotion in people's feeds, hence more responses and information that will help in your decision.) Many people who have adopted older dogs report that it wasn't easy but that they're very glad they did it and don't regret it. This isn't conclusive evidence that it's a good idea, but it's enough to update your prior belief. Since your confidence was already 50 percent, without using any formulas you feel comfortable that your subjective probability is now solidly above 50 percent and high enough to take the plunge.

With a new dog on her way to your home, you'll need a new dog bed. An extra-large one, for this 120-pound Saint. You search on Amazon and are careful to switch the Sort By setting from Featured to Average Customer Review. When the results come up, you look for the Sponsored label on the top results so you can scroll right past them. You find a big comfy bed that dog owners seem to love and that's reasonably priced.

The rescue organization said your girl is a purebred Saint Bernard, but you're not so sure. When you asked them about it, they admitted they don't know her origin story and are just guessing based on her looks. They mentioned that you can send out a saliva sample for a doggie DNA test, but they cautioned that these aren't too reliable and that different companies often report different results. You decide to give it a try, and you send samples to two companies so you can compare their findings.

Next on your list is choosing a veterinarian. There are a few in town, so you hatch a plan. You rank them based on customer reviews, price, and proximity to your house. You weight these factors according to what matters most to you, then use a weighted sum to combine these rankings and select the vet that's the best overall choice for you.

Your first visit to the vet mostly goes well, except they find some initial signs of neuropathy. They explain that four out of five times it stays at this harmless stage and never progresses into an issue. But in the remaining 20 percent of cases, it steadily worsens—and when it does, the dog's hind legs can give out, making a doggie wheelchair necessary. Fortunately, there is a medication that prevents the condition from progressing. Unfortunately, it's quite expensive. You're not sure if you should spend so much money on something that has an 80 percent chance of being unnecessary.

You decide to do a Romerian cost-benefit analysis. You imagine your dog in a wheelchair from this condition and ballpark how much you'd be willing to pay to cure her. This is the cost of the harm. You multiply this by 20 percent to get the expected loss. Then you double as Romer suggests to get the cost of the risk. The number you arrive at is considerably less than the exorbitant price of this medicine, so you decide to roll the dice and take your chances without it. You promise your fluffy friend that all the money you save from the medicine will go to giving her a great quality of life. This includes trips to the beach, the most delicious treats, and a top-of-the-line wheelchair if it comes to that. You don't know if you made the right choice, but you feel better having given it careful thought. And you remember poker champ Liv Boeree's point: so many different things can go wrong in so many ways that you shouldn't beat yourself up when bad luck strikes. You're doing your best to handle risk rationally, and you take solace in that fact, no matter how things turn out.

The DNA test results come back: one says 80 percent Saint Bernard and 20 percent Great Pyrenees, the other says half Saint Bernard and half Great Pyrenees. You decide to combine these estimates using an average, giving them equal weight, so you call it 65 percent Bernard, 35 percent Pyrenees. The exact numbers don't really matter; you're just curious. But you're happy with this news, because Great Pyrenees are among the longest living of the extra-large dog breeds. You're feeling a little more hopeful about your new friend's longevity now.

Life is full of twists and turns that are difficult to navigate. There's no guarantee that math will put you on the right path. But it can help you consider your options more clearly and make your decisions more thoughtfully. I hope the lessons in this book show you how.

Acknowledgments

First, I'd like to thank my agent, Luba Ostashevsky, for seeing the potential writer in me before I saw it myself. You took me on as a client when there was little reason to do so, and you patiently discouraged me from one bad idea after another until we hit upon the kernel that grew into this book. Next, I wish to thank my editor at Riverhead, Courtney Young, for countless insightful suggestions that consistently pushed the book in the right directions. I'm embarrassed to imagine what this book would have been like without your wise guidance. My UK editor, Alex Christofi, also provided crucial feedback that helped tie the book together. Heartfelt gratitude is owed to Paul Conner, my copyeditor, whose numerous corrections and improvements covered the gamut from proper translations of Tolstoy and Einstein to nuances of quantum mechanics and English grammar.

Your feedback didn't just improve the book tremendously; it taught me how to be a better writer. The production and publicity teams at Riverhead have been a delight to work with, and any success the book finds is owed more to them than to me.

Beyond my professional contacts in the publishing world, I would like to thank several individuals. Paul Romer graciously shared many ideas and details that found their way into the book. Your friendship and praise gave me the confidence and energy I needed both to start this book and to finish it. Molly Hickman and Julia Angwin took time out of their busy schedules to let me interview them for the book. My brother, Jeffrey Giansiracusa, read every word of a draft and offered invaluable feedback. My wife, Emily Frey, served as my unofficial editor, kindheartedly lending me her compositional and stylistic gifts. You suggested several chapter reorganizations, many passage clarifications, and countless rewordings—all of which I accepted, and all of which significantly improved the book. And my sister-in-law, Tricia Frey, lent her graphic design expertise to help perfect the cover art.

Since this is an educational book about math, I must thank all the math educators in my life who helped me learn the discipline and get to where I am today. Special thanks go to Jim Morrow, the first math teacher I had in college and the first person who made me believe I could become a mathematician. To Gerald Seidler, who never let me give up on myself. To Dan Abramovich, who guided me through graduate school and who I still count as a close mentor and friend.

ACKNOWLEDGMENTS

Finally, to my father, who taught me the logic of grammar and the excitement of math. To my mother, who taught me everything else. And again to my brother, who has been leading the way and holding my hand for every step of my life's journey, including writing this book.

Notes

1. A DAY IN THE LIFE OF A NUMBER

2 **There's an entire industry:** Jon Keegan and Alfred Ng, "Who Is Collecting Data from Your Car?," *Markup*, July 27, 2022, https://themarkup.org/the-breakdown/2022/07/27/who-is-collecting-data-from-your-car.

3 **twenty-five gigabytes of data:** Geoffrey Fowler, "What Does Your Car Know About You? We Hacked a Chevy to Find Out," *Washington Post*, December 17, 2019, https://www.washingtonpost.com/technology/2019/12/17/what-does-your-car-know-about-you-we-hacked-chevy-find-out.

3 **English-language Wikipedia articles:** "Wikipedia: Size of Wikipedia," Wikipedia, accessed July 6, 2024, https://en.wikipedia.org/wiki/Wikipedia:Size_of_Wikipedia.

3 **might not be long:** Dunkinmydonuts1, "Jeep puts ads on the fucking screen in my car," Reddit, accessed July 6, 2024, https://www.reddit.com/r/assholedesign/comments/13kgqg9/jeep_puts_ads_on_the_fucking_screen_in_my_car.

3 **has over thirty-five petabytes:** Jon Keegan, "Forget Milk and Eggs: Supermarkets Are Having a Fire Sale on Data About You," *Markup*, February 16, 2023, https://themarkup.org/privacy/2023/02/16/forget-milk-and-eggs-supermarkets-are-having-a-fire-sale-on-data-about-you.

4 **Tesco and Sainsbury's sell:** Mark Duell, "Is Your Loyalty Card REALLY a Good Deal? How Price of Biscuits, Toilet Roll and Chocolate Are Hiked Days Before Offering Discounts—as Supermarkets Make £300m from Selling Customer Data amid a Surge in New Sign-Ups," *Daily Mail*, January 8, 2024, https://www.dailymail.co.uk/news/article-12937905/Tesco-Sainsburys-supermarket-loyalty-cards-price-war.html.

5 **ten largest private employers:** Jodi Kantor and Arya Sundaram, "The Rise of the Worker Productivity Score," *New York Times*, August 14, 2022, https://www.nytimes.com/interactive/2022/08/14/business/worker-productivity-tracking.html.

5 **investment management firm:** Adam Grant, *Originals: How Non-Conformists Move the World* (New York: Viking, 2016), chapter 7.

6 **hospice chaplain's employer:** Kantor and Sundaram, "Worker Productivity Score."

6 **demand for employee monitoring:** Danielle Abril, "Your Boss Can Monitor Your Activities Without Special Software," *Washington Post*, October 7, 2022, https://www.washingtonpost.com/technology/2022/10/07/work-app-surveillance.

6 **coaches businesses on:** Thomas Sutherland, "Using AI in Employee Monitoring in the UK," Legal Vision, May 7, 2024, https://legalvision.co.uk/employment/pros-and-cons-to-using-ai-in-employee-monitoring.

7 **said that its troves:** Tara García Mathewson, "He Wanted Privacy. His College Gave Him None," *Markup*, November 30, 2023, https://themarkup.org/machine-learning/2023/11/30/he-wanted-privacy-his-college-gave-him-none.

7 **To some government agencies:** Dylan Matthews, "The Tricky Business of Putting a Dollar Value on a Human Life," *Vox*, December 22, 2022, https://www.vox.com/future-perfect/23449849/social-cost-carbon-value-statistical-life-epa.

2. RETHINK RANKINGS

13 **received a distressing email:** My wife, a professor at Brandeis, showed me this email.

14 **An official response:** Alan Blinder, "With a New Formula, U.S. News Rankings Boost Some State Universities," *New York Times*, September 18, 2023, https://www.nytimes.com/2023/09/18/us/us-news-college-ranking.html.

14 **cheered on this accomplishment:** *Fresno Bee* Editorial Board, "Go Bulldogs! Latest National Ranking Highlights Fresno State's Academic Performance," *Fresno Bee*, September 20, 2023, https://www.fresnobee.com/opinion/editorials/article279533784.html.

15 *U.S. News'* **process:** This material is drawn from Robert Morse and Eric Brooks, "How *U.S. News* Calculated the 2024 Best Colleges Rankings," *U.S. News*, September 17, 2023, https://www.usnews.com/education/best-colleges/articles/how-us-news-calculated-the-rankings; and Robert Morse and Eric Brooks, "A More Detailed Look at the Ranking Factors," *U.S. News*, September 17, 2023, https://www.usnews.com/education/best-colleges/articles/ranking-criteria-and-weights.

18 **average graduation rate:** These statistics were computed from "Use the Data," National Center for Education Statistics, accessed July 6, 2024, https://nces.ed.gov/ipeds/use-the-data.

19 *Times Higher Education* **in the UK:** "World University Rankings 2024: Methodology," *Times Higher Education*, September 20, 2023, https://www.timeshighereducation.com/world-university-rankings/world-university-rankings-2024-methodology.

22 **College Navigator website:** "Find the Right College for You," National Center for Education Statistics, https://nces.ed.gov/collegenavigator.

29 **posted on his webpage:** Michael Thaddeus, "An Investigation of the Facts Behind Columbia's *U.S. News* Ranking," Michael Thaddeus's Columbia University homepage, revised March 2022, https://www.math.columbia.edu/~thaddeus/ranking/investigation.html.

29 **looking into the matter:** Gloria Oladipo, "*US News* College Ranking Under Scrutiny After Ivy League University Plummets," *Guardian*, September 13, 2022, https://www.theguardian.com/us-news/2022/sep/13/us-news-college-ranking-controversy-columbia-university.

31 **a damning report:** "Investigation of Data Irregularities in *Doing Business 2018* and *Doing Business 2020*," World Bank, September 15, 2021, https://thedocs.worldbank.org/en/doc/84a922cc9273b7b120d49ad3b9e9d3f9-0090012021/original/DB-Investigation-Findings-and-Report-to-the-Board-of-Executive-Directors-September-15-2021.pdf.

32 *Doing Business* **rankings:** "Chapter 6: Ease of Doing Business Score and Ease of Doing Business Ranking," World Bank, https://openknowledge.worldbank.org/server/api/core/bitstreams/2db10c90-db63-5746-a4e3-1bfe264dda7c/content.

34 **debacle put an end:** Andrea Shalal and David Lawder, "IMF Chief Called Out over Pressure to Favor China While at World Bank," *Reuters*, September 17, 2021, https://www.reuters.com/business/sustainable-business/world-bank-kills-business-climate-report-after-ethics-probe-cites-undue-pressure-2021-09-16.

35 **held eight meetings:** Alan Rappeport, "Kristalina Georgieva Will Remain Managing Director of the I.M.F., Its Board Says," *New York Times*, October 11, 2021, https://www.nytimes.com/2021/10/11/business/kristalina-georgieva-imf.html.

37 **weighted sum of prices:** See Nasiha Salwati and David Wessel, "How Does the Government Measure Inflation?," Brookings Institution, June 28, 2021, https://www.brookings.edu/articles/how-does-the-government-measure-inflation; and "Consumer Price Index," US Bureau of Labor Statistics, accessed July 6, 2024, https://www.bls.gov/cpi/tables/relative-importance/home.htm.

38 **FICO revealed that:** "What's in My FICO Scores?," myFICO, accessed July 6, 2024, https://www.myfico.com/credit-education/whats-in-your-credit-score.

39 **Some experts have argued:** James DePorre, "Ignore the Misleading Dow Jones Industrial Average," *The Street*, September 7, 2018, https://pro.thestreet.com/investing/ignore-misleading-dow-jones-industrial-average.

39 **Asher and Lyric Fergusson publish:** Asher Fergusson and Lyric Fergusson,

"The 203 Worst (& Safest) Countries for LGBTQ+ Travel in 2023," updated June 5, 2023, https://www.asherfergusson.com/lgbtq-travel-safety.

3. MAKE PREDICTIONS LIKE A PRO

44 **1906 Plymouth exhibition:** Francis Galton, "Vox Populi," *Nature* 75, no. 1949 (March 1907): 450–51, https://www.nature.com/articles/075450a0.

45 **he apparently enjoyed:** Stephanie Clifford, "Math Whiz Finds Fame by Calling It for Obama," *New York Times*, October 10, 2008, https://www.nytimes.com/2008/11/10/business/worldbusiness/10iht-silver.1.17680229.html.

52 **71 percent chance of winning:** "Who Will Win the Presidency?," *FiveThirty-Eight*, November 8, 2016, https://projects.fivethirtyeight.com/2016-election-forecast.

54 **That year** *The Economist*: "How *The Economist* Presidential Forecast Works," *Economist*, November 3, 2020, https://projects.economist.com/us-2020-forecast/president/how-this-works.

54 **columnist Perry Bacon Jr.:** Perry Bacon Jr., "Polls Are Useful. They Just Can't Predict Elections in Swing States," *Washington Post*, September 27, 2022, https://www.washingtonpost.com/opinions/2022/09/27/polls-accuracy-midterm-elections-perry-bacon.

55 **President Trump released guidelines:** "President Donald J. Trump Announces Guidelines for Opening Up America Again," Trump White House Archives, April 16, 2020, https://trumpwhitehouse.archives.gov/briefings-statements/president-donald-j-trump-announces-guidelines-opening-america.

55 **cited the pandemic predictions:** Adam Cancryn, "How Overly Optimistic Modeling Distorted Trump Team's Coronavirus Response," *Politico*, April 24, 2020, https://www.politico.com/news/2020/04/24/trump-coronavirus-model-207582.

56 **What they did:** "COVID-19," Institute for Health Metrics and Evaluation, accessed July 6, 2024, https://www.healthdata.org/research-analysis/diseases-injuries/covid-our-approach.

57 **ridiculed the epidemiological approach:** Cancryn, "Overly Optimistic Modeling."

57 **The epidemiologists struck back:** Cancryn, "Overly Optimistic Modeling."

58 **added an epidemiologist:** Sharon Begley, "Influential Covid-19 Model Uses Flawed Methods and Shouldn't Guide U.S. Policies, Critics Say," CNBC, April 17, 2020, https://www.cnbc.com/2020/04/17/influential-covid-19-model-uses-flawed-methods-and-shouldnt-guide-us-policies-critics-say.html.

58 **They combined multiple:** "Ensemble Model," COVID-19 Forecast Hub, accessed July 6, 2024, https://covid19forecasthub.org/doc/ensemble.

61 **rule along ideological lines:** Amelia Thomson-DeVeaux and Laura Bronner, "The Supreme Court's Partisan Divide Hasn't Been This Sharp in Generations," *FiveThirtyEight*, July 5, 2022, https://fivethirtyeight.com/features/the-supreme-courts-partisan-divide-hasnt-been-this-sharp-in-generations.

64 **"an absolutely obscene margin":** "Mantic Monday 3/14/22," *Astral Codex Ten*, March 14, 2022, https://www.astralcodexten.com/p/mantic-monday -31422.

64 **offered some advice:** Molly Hickman, "How Not to Predict the Future," *Asterisk*, March 2024, https://asteriskmag.com/issues/05/how-not-to-predict -the-future.

64 **When I asked Hickman:** Molly Hickman, interview by the author, July 2024.

4. WHAT TO EXPECT WHEN YOU'RE EXPECTING VALUE

67 **Paid celebrity endorsements:** Ari Redbord, "Tom Brady and Other A-Listers Sued for Fumbling FTX Endorsements," *Forbes*, February 2, 2023, https://www.forbes.com/sites/ariredbord/2023/02/01/tom-brady-and -other-a-listers-fumble-ftx-endorsements-but-will-they-be-held-liable.

67 **A prosecutor described:** "Statement of U.S. Attorney Damian Williams on the Conviction of Samuel Bankman-Fried," Department of Justice, November 2, 2023, https://www.justice.gov/usao-sdny/pr/statement-us-attorney -damian-williams-conviction-samuel-bankman-fried.

67 **holding back tears:** Elizabeth Lopatto, "Sam Bankman-Fried Gambled on a Trial and His Parents Lost," *Verge*, November 2, 2023, https://www.theverge .com/2023/11/2/23944485/sam-bankman-fried-guilty-verdict-parents.

68 **In it, Lewis writes:** Michael Lewis, "Play It Again, Sam," *Washington Post*, October 1, 2023, https://www.washingtonpost.com/opinions/interactive/2023 /michael-lewis-sam-bankman-fried-ftx-crypto.

69 **wasn't opposed to quantum:** George Musser, "What Einstein Really Thought about Quantum Mechanics," *Scientific American*, September 1, 2015, https://www.scientificamerican.com/article/what-einstein-really-thought -about-quantum-mechanics.

76 **once said of roulette:** Stephen Hawking, "The Future of the Universe," *Engineering & Science*, Fall 1991, https://calteches.library.caltech.edu/3684/1 /Universe.pdf.

77 **In the early morning:** Kit Chellel, "The Gambler Who Beat Roulette," *Bloomberg*, April 6, 2023, https://www.bloomberg.com/features/2023-how-to -beat-roulette-gambler-figures-it-out/.

84 **headlines across Texas:** Dave Lieber, "Syndicates Are Spending Millions of Dollars on Texas' Lottery to Beat out Everyone Else," *Dallas Morning News*, April 27, 2023, https://www.dallasnews.com/news/watchdog/2023 /04/27/bought-a-ticket-for-lotto-texas-recent-95-million-jackpot-you-had -heavy-competition.

90 **exchanged text messages:** Kelsey Piper, "Sam Bankman-Fried Tries to Explain Himself," *Vox*, November 16, 2022, https://www.vox.com/future-perfect /23462333/sam-bankman-fried-ftx-cryptocurrency-effective-altruism -crypto-bahamas-philanthropy.

92 **Warren Buffett once said:** Tren Griffin, "A Dozen Things I've Learned from Charlie Munger About Risk, Part 1," *Columbia University Press Blog*,

September 17, 2015, https://cupblog.org/2015/09/17/a-dozen-things-ive-learned-from-charlie-munger-about-risk-part-1.

5. HOW TO HANDLE RISK

95 *The New Yorker* ran: Kathryn Schulz, "The Really Big One," *New Yorker*, July 13, 2015.

96 link between the vaccine: "New Study Updates Evidence on Rare Heart Condition After Covid Vaccination," BMJ Group, July 15, 2022, https://www.bmj.com/company/newsroom/new-study-updates-evidence-on-rare-heart-condition-after-covid-vaccination.

97 five highest-paid players: This and the following basketball salary statistics are from "How Much Do NBA Players Make? Average Salary from 1990–2022," The Hoops Geek, July 2, 2022, https://www.thehoopsgeek.com/average-nba-salary.

98 novelist Mary Adkins: Mary Adkins, "How Much Can an Author Expect to Make on Their Book?," blog post, accessed July 6, 2024, https://maryadkinswriter.com/blog/how-much-do-authors-make.

98 unlucky 6.5 percent or so: Claire Hastie et al., "True Prevalence of Long-COVID in a Nationwide, Population Cohort Study," *Nature Communications* 14, no. 7892 (November 2023), https://www.nature.com/articles/s41467-023-43661-w.

99 Throughout the last millennium: Aaron O'Neill, "Global Life Expectancy from Birth in Selected Regions 1000–2020," *Statista*, July 4, 2024, https://www.statista.com/statistics/1303775/global-life-expectancy-by-region-country-historical.

99 In medieval England: Sharon DeWitte, "Old Age Isn't a Modern Phenomenon—Many People Lived Long Enough to Grow Old in the Olden Days, Too," University of South Carolina, August 10, 2022, https://sc.edu/uofsc/posts/2022/08/conversation-old-age-is-not-a-modern-phenomenon.php.

101 released the sixth: Sophie Boehm and Clea Schumer, "10 Big Findings from the 2023 IPCC Report on Climate Change," World Resources Institute, March 20, 2023, https://www.wri.org/insights/2023-ipcc-ar6-synthesis-report-climate-change-findings.

103 around the tenth century: Brian Atwater et al., "Earthquake Recurrence Inferred from Paleoseismology," *Developments in Quaternary Science* 1 (2003): 331–50, https://www.sciencedirect.com/science/article/abs/pii/S1571086603010157.

107 it would be worth: You can do calculations like this with the following web app, which also explains the math behind the calculation: "Compound Annual Growth Rate (Annualized Return)," Moneychimp, accessed July 6, 2024, http://www.moneychimp.com/features/market_cagr.htm.

108 "By periodically investing": The Buffett quotes and the rest of the Buffett material in this chapter are drawn from Robert G. Hagstrom, *The Warren Buffett Way*, 3rd ed. (Hoboken, NJ: Wiley, 2014).

112 **their remarkable findings:** Zack Beauchamp, "This Study Tried to Improve Our Ability to Predict Major Geopolitical Events. It Worked," *Vox*, August 21, 2015, https://www.vox.com/2015/8/20/9179657/tetlock -forecasting.

112 **her collaborators revisited:** Matthew Killingsworth, Daniel Kahneman, and Barbara Mellers, "Income and Emotional Well-Being: A Conflict Resolved," *Proceedings of the National Academy of Sciences* 120, no. 10 (March 2023), https://www.pnas.org/doi/10.1073/pnas.2208661120.

116 **Romer had advocated:** Isaac Chotiner, "Paul Romer's Case for Nationwide Coronavirus Testing," *New Yorker*, May 3, 2020, https://www.newyorker .com/news/q-and-a/paul-romer-on-how-to-survive-the-chaos-of-the -coronavirus.

117 **end of this chapter, has written:** Liv Boeree, "What the World Needs Now: Lessons from a Poker Player," *Nature* 582, no. 7813 (June 2020): 480, https:// www.nature.com/articles/d41586-020-01840-5.

117 **Romer shared with me:** Paul Romer, interview by the author, June 2022.

120 **"Imagine I'm doing":** Romer, interview.

121 **around 6.5 percent of these:** Estimates vary widely on this number, from around 3 percent to 20 percent. Here I'm using the 6.5 percent figure cited earlier in the chapter.

123 **Caitlin Clark, the unparalleled:** For a mathy take on what makes her so great, see Ben Pickman, "What Makes Caitlin Clark the Best Shooter in College Basketball? The Physics Behind Her Shot," *Athletic*, March 19, 2024, https://theathletic.com/5351405/2024/03/19/caitlin-clark-shot-mechanics -ncaa-tournament.

125 **Haghani and a colleague:** Victor Haghani and Richard Dewey, "Rational Decision-Making Under Uncertainty: Observed Betting Patterns on a Biased Coin," arXiv preprint, October 19, 2016, https://arxiv.org/pdf/1701.01427.pdf.

130 **he finished eighty-seventh:** Nate Silver (@NateSilver538), "Just made a huge, correct preflop all in call in the prior hand," July 12, 2023, https://twitter .com/NateSilver538/status/1679289459857293312.

130 **Boeree explained that chess:** Liv Boeree, "Episode 6: Liv Boeree on Poker, Aliens, and Thinking in Probabilities," interview by Sean Carroll, *Mindscape*, July 23, 2018, https://www.preposterousuniverse.com/podcast/2018 /07/23/episode-6-liv-boeree-on-poker-aliens-and-thinking-in-probabilities.

131 **in a TED Talk:** Liv Boeree, "A Number Speaks a Thousand Words," TEDxManchester, February 2018, https://www.ted.com/talks/liv_boeree_a _number_speaks_a_thousand_words.

6. A TOOL FOR THINKING MORE CLEARLY

133 **she later wrote:** Liv Boeree, "How an 18th-Century Priest Gave Us the Tools to Make Better Decisions," *Vox*, November 30, 2018, https://www.vox .com/future-perfect/2018/11/30/18096751/bayes-theorem-rule-rationality -reason.

138 **"Unless we know":** Thomas Bayes, *Divine Benevolence: Or, an Attempt to Prove That the Principal End of the Divine Providence and Government Is the Happiness of His Creatures* (London: John Noon, 1731).

139 **an extended introduction:** Thomas Bayes, "An Essay Towards Solving a Problem in the Doctrine of Chances. By the Late Rev. Mr. Bayes, F. R. S. Communicated by Mr. Price, in a Letter to John Canton, A. M. F. R. S," *Philosophical Transactions* 53 (December 1763), https://royalsocietypublishing .org/doi/10.1098/rstl.1763.0053.

145 **"We knew that we could":** Alan M. Dershowitz, *Reasonable Doubts: The O.J. Simpson Case and the Criminal Justice System* (New York: Simon & Schuster, 1996), 101.

145 **in a brief letter:** Jack Good, "When Batterer Becomes Murderer," *Nature* 381, no. 481 (June 1996), https://www.nature.com/articles/381481a0.

146 **Good had an illustrious:** Dan van der Vat, "Obituary: Jack Good," *Guardian*, April 28, 2009, https://www.theguardian.com/science/2009/apr/29/jack -good-codebreaker-obituary.

152 **debate about this:** Dobromir Rahnev, "The Bayesian Brain: What Is It and Do Humans Have It?," *Behavioral and Brain Sciences* 42 (2019), https://doi .org/10.1017/S0140525X19001377.

152 **whether she uses Bayes's:** Liv Boeree, "Episode 6: Liv Boeree on Poker, Aliens, and Thinking in Probabilities," interview by Sean Carroll, *Mindscape*, July 23, 2018, https://www.preposterousuniverse.com/podcast/2018/07/23 /episode-6-liv-boeree-on-poker-aliens-and-thinking-in-probabilities.

7. THE SECRET SOCIAL MEDIA FORMULA

157 **The internet was flummoxed:** See Emerald Pellot, "Twitter May Have Figured Out the Answer to That Viral 'Nursing Assistant' TikTok Riddle: 'What Did She Even MEAN?'," *Yahoo! Life*, January 12, 2023, https://www .yahoo.com/lifestyle/twitter-may-have-figured-out-the-answer-to-that -viral-nursing-assistant-tik-tok-riddle-what-did-she-even-mean -204440072.html; and Jacob Seitz, "'Nurning? Annintant?': Everyone Is Baffled by This 26yo Old Nursing Assistant's Cryptic TikTok Riddle," *Daily Dot*, December 14, 2022, https://www.dailydot.com/debug/nursing-assistant -tiktok-viral.

159 **2014 research paper published:** Adam Kramer, Jamie Guillory, and Jeffrey Hancock, "Experimental Evidence of Massive-Scale Emotional Contagion Through Social Networks," *Proceedings of the National Academy of Sciences* 111, no. 24 (June 2014): 8788–90, https://www.pnas.org/doi/full/10.1073 /pnas.1320040111.

160 *Atlantic* **ran a story:** Robinson Meyer, "Everything We Know About Facebook's Secret Mood-Manipulation Experiment," *Atlantic*, June 28, 2014, https://www.theatlantic.com/technology/archive/2014/06/everything-we -know-about-facebooks-secret-mood-manipulation-experiment/373648.

160 **personal data from eighty-seven:** Olivia Solon, "Facebook Says Cambridge Analytica May Have Gained 37M More Users' Data," *Guardian*, April 4, 2018, https://www.theguardian.com/technology/2018/apr/04/facebook-cambridge -analytica-user-data-latest-more-than-thought.

160 **widely shared article titled:** Adrienne LaFrance, "Facebook Is a Dooms- day Machine," *Atlantic*, December 15, 2020, https://www.theatlantic.com /technology/archive/2020/12/facebook-doomsday-machine/617384.

161 **a lengthy blog post:** Nick Clegg, "You and the Algorithm: It Takes Two to Tango," Medium, March 31, 2021, https://nickclegg.medium.com/you-and -the-algorithm-it-takes-two-to-tango-7722b19aa1c2.

162 **the company's engineering blog:** Akos Lada, Meihong Wang, and Tak Yan, "How Machine Learning Powers Facebook's News Feed Ranking Algo- rithm," Meta, January 26, 2021, https://engineering.fb.com/2021/01/26/core -infra/news-feed-ranking.

163 **the biggest scandals:** Jeff Horwitz et al., "The Facebook Files," *Wall Street Journal*, September–December 2021, https://www.wsj.com/articles /the-facebook-files-11631713039.

169 **Interestingly, the angry emoji:** Jeremy Merrill and Will Oremus, "Five Points for Anger, One for a 'Like': How Facebook's Formula Fostered Rage and Misinformation," *Washington Post*, October 26, 2021, https://www.washing tonpost.com/technology/2021/10/26/facebook-angry-emoji-algorithm.

169 **Facebook recently revealed that:** See "Our Approach to Explaining Ranking," Meta, December 31, 2023, https://transparency.fb.com/features /explaining-ranking; and "Our Approach to Facebook Feed Ranking," Meta, November 28, 2023, https://transparency.fb.com/features/ranking-and -content.

171 ***Times* hunted for answers:** Ben Smith, "How TikTok Reads Your Mind," *New York Times*, December 5, 2021, https://www.nytimes.com/2021/12/05 /business/media/tiktok-algorithm.html.

175 **A resounding 83 percent said:** Mitchell Clark, "Twitter Takes Its Algorithm 'Open-Source,' as Elon Musk Promised," *Verge*, March 31, 2023, https:// www.theverge.com/2023/3/31/23664849/twitter-releases-algorithm-musk -open-source.

175 **Musk prefaced the code:** Kyle Wiggers, "Twitter Reveals Some of Its Source Code, Including Its Recommendation Algorithm," *Tech Crunch*, March 31, 2023, https://techcrunch.com/2023/03/31/twitter-reveals-some-of-its-source -code-including-its-recommendation-algorithm.

175 **rather adolescent backstory:** Kari Paul, "Elon Musk Reportedly Forced Twitter Algorithm to Boost His Tweets After Super Bowl Flop," *Guardian*, February 15, 2023, https://www.theguardian.com/technology/2023/feb/15 /elon-musk-changes-twitter-algorithm-super-bowl-slump-report.

176 **thanks to the code release:** Igor Brigadir and Vicki Boykis, "Awesome Twitter Algo," GitHub, accessed July 6, 2024, https://github.com/igorbrigadir/awesome -twitter-algo.

176 **In Twitter's formula:** In the actual code, all these numbers are half what I wrote here. I doubled them for convenience, since rescaling all the weights doesn't have any impact.

8. STANDING UP TO THE TECH TITANS

181 **largest companies by market:** Adam Hayes, Cierra Murry, and Ryan Eichler, "Biggest Companies in the World by Market Cap," *Investopedia*, July 3, 2024, https://www.investopedia.com/biggest-companies-in-the-world-by -market-cap-5212784.

184 **One ex-employee said:** Loren Baker, "Amazon's Search Engine Ranking Algorithm: What Marketers Need to Know," *Search Engine Journal*, August 14, 2018, https://www.searchenginejournal.com/amazon-search-engine-ranking -algorithm-explained/265173.

184 **investigation by *The Markup*:** Adrianne Jeffries and Leon Yin, "Amazon Puts Its Own 'Brands' First," *Markup*, October 14, 2021, https://themarkup .org/amazons-advantage/2021/10/14/amazon-puts-its-own-brands-first -above-better-rated-products.

185 **Angwin told me:** Julia Angwin, interview by the author, June 2024.

186 **The CEO of Allbirds:** Joey Zwillinger, "Dear Mr. Bezos," Medium, November 25, 2019, https://joeyzwillinger.medium.com/dear-mr-bezos-e691f6d 6d705.

187 **One study found:** Rory Van Loo and Nikita Aggarwal, "Amazon's Pricing Paradox," Boston University School of Law, October 5, 2023, https:// scholarship.law.bu.edu/faculty_scholarship/3645.

188 **A study in 2021:** Geoffrey Fowler, "It's Not Your Imagination: Shopping on Amazon Has Gotten Worse," *Washington Post*, November 23, 2022, https:// www.washingtonpost.com/technology/interactive/2022/amazon-shopping -ads/.

188 **Doctorow has a clever:** Cory Doctorow, "View a SKU," Medium, July 10, 2022, https://doctorow.medium.com/view-a-sku-32721d623aee.

189 **slick web browser plug-in:** Library Extension, https://www.libraryex tension.com.

189 **estimated to reach:** Nikki Main, "Amazon Is Pocketing Half of Retailers' Sales," *Gizmodo*, February 13, 2023, https://gizmodo.com/amazon-amazon -prime-marketplace-online-shopping-1850109105.

190 **American laws don't adequately:** The situation is better in the EU, thanks to the Digital Markets Act; see "About the Digital Markets Act," European Commission, https://digital-markets-act.ec.europa.eu/about-dma_en.

190 **nearly 80 percent of Google's:** Tiago Bianchi, "Google: Annual Advertising Revenue 2001–2023," *Statista*, May 22, 2024, https://www.statista.com /statistics/266249/advertising-revenue-of-google.

190 **accounts for the majority:** "World Advertising Industry Spend & Revenue Statistics," Voicebooking, accessed July 6, 2024, https://www.voicebooking .com/en/world-advertising-industry-spend-revenue-statistics.

190 **Next on the list:** Ronan Shields, "Here Are the 2022 Global Media Rankings by Ad Spend: Google, Facebook Remain Dominant—Alibaba, ByteDance in the Mix," *Digiday*, December 13, 2022, https://digiday.com/media/the-rundown-here-are-the-2022-global-media-rankings-by-ad-spend-google-facebook-remain-dominate-alibaba-bytedance-in-the-mix.

191 **nearly 80 percent of Amazon's:** "Amazon Advertising Report 2024," Jungle Scout, accessed July 6, 2024, https://www.junglescout.com/resources/reports/2024-amazon-advertising-report.

191 *Washington Post* **technology columnist:** Geoffrey Fowler, "How Does Google's Monopoly Hurt You? Try These Searches," *Washington Post*, October 20, 2020, https://www.washingtonpost.com/technology/2020/10/19/google-search-results-monopoly.

192 **You can use a minus sign:** For a fuller list of options than what I present here, see Daniel Russell, "Advanced Search Operators," February 8, 2024, https://docs.google.com/document/d/1ydVaJJeL1EYbWtlfj9TPfBTE5IBADkQfZrQaBZxqXGs/edit.

194 **an op-ed in** *Slate***:** Noah Giansiracusa, "Google Needs to Defund Misinformation," *Slate*, November 18, 2021, https://slate.com/technology/2021/11/google-ads-misinformation-defunding-artificial-intelligence.html.

194 **federal antitrust case:** David McCabe and Nico Grant, "U.S. Accuses Google of Abusing Monopoly in Ad Technology," *New York Times*, January 24, 2023, https://www.nytimes.com/2023/01/24/technology/google-ads-lawsuit.html.

195 **One report estimated:** Matt Skibinski, "Special Report: Top Brands Are Sending $2.6 Billion to Misinformation Websites Each Year," NewsGuard, accessed July 6, 2024, https://www.newsguardtech.com/special-reports/brands-send-billions-to-misinformation-websites-newsguard-comscore-report.

195 **other investigations found:** "The Quarter Billion Dollar Question: How Is Disinformation Gaming Ad Tech?," Global Disinformation Index, September 1, 2019, https://www.disinformationindex.org/research/2019-9-1-the-quarter-billion-dollar-question-how-is-disinformation-gaming-ad-tech.

198 **However, some experts:** Ethan Zuckerman, "The Internet's Original Sin," *Atlantic*, August 14, 2014, https://www.theatlantic.com/technology/archive/2014/08/advertising-is-the-internets-original-sin/376041.

198 **Instagram, for instance:** Mansoor Iqbal, "Instagram Revenue and Usage Statistics (2024)," Business of Apps, April 18, 2024, https://www.businessofapps.com/data/instagram-statistics.

199 **decade later, nearly 95 percent:** Simon Kemp, "Digital 2023: The Philippines," Datareportal, February 9, 2023, https://datareportal.com/reports/digital-2023-philippines.

199 **carried a tremendous price:** Jake Swearingen, "Facebook Used the Philippines to Test Free Internet. Then a Dictator Was Elected," *New York*, September 4, 2018, https://nymag.com/intelligencer/2018/09/how-facebooks-free-internet-helped-elect-a-dictator.html.

199 **killed thousands of people:** "How Many People Have Been Killed in Ro-

drigo Duterte's War on Drugs?," *Economist*, November 22, 2021, https://
www.economist.com/graphic-detail/2021/11/22/how-many-people-have
-been-killed-in-rodrigo-dutertes-war-on-drugs.

199 **Kara Swisher interviewing Nobel:** Maria Ressa, "Will Maria Ressa's
Nobel Peace Prize Force Mark Zuckerberg to Wake Up?," interview by Kara
Swisher, *Sway* (podcast), *New York Times*, October 21, 2021, https://www
.nytimes.com/2021/10/21/opinion/sway-kara-swisher-maria-ressa.html.

200 **spaces on the internet:** "The Digital Human, Series 29," BBC, June 26,
2023, https://www.bbc.co.uk/sounds/play/m001n8f5.

201 **recent viral essay:** Cory Doctorow, "The 'Enshittification' of TikTok,"
Wired, January 23, 2023, https://www.wired.com/story/tiktok-platforms-cory
-doctorow.

201 **Word of the Year:** "2023 Word of the Year Is 'Enshittification,'" American
Dialect Society, January 5, 2024, https://americandialect.org/2023-word-of
-the-year-is-enshittification.

202 **Federal Trade Commission, said:** David Dayen, "One Person One Price,"
American Prospect, June 4, 2024, https://prospect.org/economy/2024-06-04
-one-person-one-price.

204 *New York Times* **op-ed:** Paul Romer, "A Tax That Could Fix Big Tech," *New
York Times*, May 6, 2019, https://www.nytimes.com/2019/05/06/opinion/tax
-facebook-google.html.

204 **a 2021 blog post:** Paul Romer, "Taxing Digital Advertising," May 17, 2021,
https://adtax.paulromer.net.

205 **ban targeted advertising:** Banning targeted advertising doesn't mean banning online advertising. In all the proposals I've seen, contextual advertising
would still be allowed. This is where the ads are selected based on the content
on the webpage, rather than the personal data of the visitor. For instance, a
basketball site would be able to host ads for sports equipment, knowing visi-
tors to the site might be interested in this. Contextual ads can still be selected
algorithmically, but the algorithm would use website data rather than user
data. This eliminates the incentive to track and surveil users.

205 **published by the Poynter:** Lucia Walinchus, "We Need a News Utility,"
Poynter, June 16, 2022, https://www.poynter.org/commentary/2022/national
-tax-support-journalism.

206 **publicly funded alternatives:** Rachel Metz, "Social Networks Are Broken.
This Man Wants to Fix Them," *MIT Technology Review*, February 9, 2018,
https://www.technologyreview.com/2018/02/09/3406/social-networks-are
-broken-this-man-wants-to-fix-them.

207 **Nobel Prize in 2018:** "Popular Information," Nobel Prize, accessed July 6,
2024, https://www.nobelprize.org/prizes/economic-sciences/2018/popular
-information.

208 **In his papers, Romer:** Paul Romer, "Are Nonconvexities Important for
Understanding Growth?," *American Economic Review* 80, no. 2 (May 1990):
97–103, https://www.jstor.org/stable/2006550.

209 **In a blog post:** Sam Altman, "What I Wish Someone Had Told Me," December 21, 2023, https://blog.samaltman.com/what-i-wish-someone-had-told-me.

210 **most powerful woman:** Lawrence Delevingne, "Meet the Most Powerful Woman in Hedge Funds," CNBC, January 8, 2015, https://www.cnbc.com /2015/01/08/meet-leda-braga-the-most-powerful-woman-in-hedge-funds .html.

210 **Braga explained:** Leda Braga, interview by the author, September 2021.

Index